# Springer Series in Materials Science

Volume 346

The Springer Series in Materials Science covers the complete spectrum of materials research and technology, including fundamental principles, physical properties, materials theory and design. Recognizing the increasing importance of materials science in future device technologies, the book titles in this series reflect the state-of-the-art in understanding and controlling the structure and properties of all important classes of materials.

Christopher D. Woodgate

# Modelling Atomic Arrangements in Multicomponent Alloys

## A Perturbative, First-Principles-Based Approach

Springer

Christopher D. Woodgate
Department of Physics
University of Warwick
Coventry, UK

ISSN 0933-033X ISSN 2196-2812 (electronic)
Springer Series in Materials Science
ISBN 978-3-031-62020-1 ISBN 978-3-031-62021-8 (eBook)
https://doi.org/10.1007/978-3-031-62021-8



This Springer imprint is published by the registered company Springer Nature Switzerland AG
The registered company address is: Gewerbestrasse 11, 6330 Cham, Switzerland

*To my parents.*

# Foreword

Many important functional materials are composed of multiple components in specific proportions. For example, conventional alloys, like steel and aluminium-based alloys, comprise one to three base metals and trace additions to stabilise the structures and tune the material properties. Commercial hard magnets are the results of trial and error processes of multicomponent additions to the principal transition metal and rare earth elements to improve performance and/or lower cost. There is also the class of high-entropy alloys (HEAs) which is attracting great interest. HEAs are disordered alloys with four or more base metals and have been found with specific strength, corrosion and wear resistance that are comparable to, or exceed that of, conventional alloys. From a scientific standpoint the motivation to explore the vast composition space inhabited by multicomponent alloys is therefore compelling.

Computational materials modelling has a key role to play in the rapid exploration and design of new functional multicomponent materials. This is challenging given the complexity of the phase spaces. At the sub-nanoscale, the same interacting electrons which bond the nuclei into their positions also determine material properties. So computational modelling needs to describe both the atomic arrangements in the various phases as well as property trends. The work described by Christopher Woodgate in this monograph addresses these challenges and lays out ground work for a powerful and flexible analysis technique for the design of new alloys. The approach is well-suited for use in collaborations with both experimental groups and other computational modellers as well as for reaching out to industry.

As with much *ab initio* materials modelling, the starting point is density functional theory (DFT). From such tenets the book sets out an approach for examining the growth of atomic short-range order and the instability of the high temperature multicomponent solid solution phases of alloys to chemical fluctuations as the temperature is lowered. This theory addresses the stability of multicomponent alloys from the free-energy change as a result of infinitesimal variations in the occupancies of the components averaged for each lattice site which are expressed in terms of 'concentration waves' and chemical polarisations. For example, a ternary $A_{\frac{1}{3}}B_{\frac{1}{3}}C_{\frac{1}{3}}$ BCC solid solution might show an instability to a concentration wave with wave-vector $(0, 0, \frac{2\pi}{a})$ ($a$ is the lattice parameter) and chemical polarisations $(0.724, -0.689, -0.033)$. This information describes an ordering into a B2 or CsCl phase with one sub-lattice rich with element $A$, the other sub-lattice $B$-rich with element $C$ distributed fairly uniformly throughout.

The free-energy change derived in the theory includes not only entropic effects but also electronic effects from rearranging charge and changing electronic structure. The theory is in essence analogous to that of density functional perturbation theory (DFPT) used to describe lattice dynamics *ab initio* and response functions for phonons where now the relevant modes here are concentration waves. The monograph explains how effective atom–atom interactions can be extracted from the *ab initio* data and used in further atomistic Monte Carlo simulations of alloy phase diagrams. Compositional information, where it exists, from experiments (*e.g.* X-ray and neutron scattering) can be analysed and new phases predicted. Applications to the Cantor-Wu CrMnFeCoNi FCC and refractory TiVNbMoTaW BCC HEAs demonstrate the approach. Moreover when compared to high-throughput DFT calculations on supercells of evaluations of energies, it is shown how the method uses orders of magnitude less computing resources. Finally the book explains how the magnetic states of the solid solution alloy phases can affect the ordering trends and how annealing in a magnetic field might be harnessed in materials synthesis. In reverse, a case study on iron-rich Fe–Ga alloys shows how the atomic arrangements affect magnetic properties.

The book is excellent and exceptionally well-written. After describing the materials science context, it carefully explains the DFT and statistical mechanics background at a level which is accessible to a wide range of researchers in computational materials science including post-graduate students. The theoretical framework is carefully developed, described and demonstrated by a sequence of case studies. The exciting scope is put across for expanding this approach to interpret spectroscopy, perform transport calculations and include effects of strong electron correlations so that, for example, multicomponent materials containing lanthanides, high-entropy oxides and carbides can be explored.

A note about the author: In October 2019 Christopher Woodgate (Chris) enrolled as one of the first cohort of students in the University of Warwick's Centre for Doctoral Training (CDT) in Modelling of Heterogeneous Systems (HetSys), which is funded by the UK Engineering and Physical Sciences Research Council, to study for a Ph.D. and follow its programme. The programme involves extensive, wide-ranging training in computational materials modelling which is integrated with the Ph.D. research work. I was Chris' Ph.D. project supervisor. Chris submitted an outstanding thesis at the end of September 2023, passed the subsequent exam in December 2023 with flying colours and has recently been awarded his Ph.D. Chris was also a fantastic member of our first cohort of HetSys CDT students, playing a great role in establishing some of the valuable outreach work it does and mentoring students in later cohorts.

Chris is in the top set of post-graduate students and early career researchers I have worked with over the years. He is very creative and contributes lots of great ideas, easy to work with and engages exceptionally well in collaborations. He gives excellent conference presentations and writes very well too, as shown by this monograph and the first author published papers it cites. It is a pleasure and honour to write this Foreword.

Prof. Julie B. Staunton
Department of Physics
University of Warwick
Coventry, UK

# Preface

This book began life as my Ph.D. thesis, titled *Atomic Arrangements in Multicomponent Alloys: First-Principles Theory, Atomistic Modelling, and Implications for Magnetic Properties*, which was submitted for examination at the University of Warwick in September 2023. The present work forms an enlargement of that earlier work, providing additional details and applications, together with a more thorough exposition of the underlying theory. I am fortunate to have been able to work on the book while remaining at the University of Warwick as a postdoctoral researcher, working in the same group at that in which I studied for my Ph.D., under the supervision of Prof. Julie B. Staunton.

The book presents the implementation and analysis of a computational modelling approach for studying the nature of atomic arrangements in multicomponent alloys. The approach is based on first-principles calculations of the electronic structure of the alloys as described within density functional theory (DFT) and the coherent potential approximation (CPA). In particular, a perturbative analysis of the internal energy of the disordered alloy as evaluated within the CPA is considered. The perturbations are inhomogeneous variations of the partial chemical occupancies of lattice sites, which are described in reciprocal space within a *concentration wave* formalism. This perturbative analysis (or 'linear-response' theory) must carefully consider the rearrangement of the electronic structure of the alloy due to an applied chemical perturbation, in a similar manner to the considerations made in density functional perturbation theory (DFPT) for obtaining lattice dynamics, *e.g.* phonons, *ab initio*. From the perturbative analysis, it is possible to infer chemical disorder-order transitions directly via an application of Landau theory. In addition, it is possible to extract an effective pair interaction describing the internal energy of the alloy that is suitable for study via, *e.g.* Monte Carlo simulations.

We begin the book in Chap. 1 by motivating the work and setting it in its wider context. Chapters 2 and 3 then describe the necessary tools from statistical physics and electronic structure respectively, culminating in an outline of the $S^{(2)}$ theory for multicomponent alloys, which is the theory underpinning our method for perturbatively analysing the energetic cost of inhomogeneous chemical fluctuations to the CPA reference medium. The subsequent four chapters present results of the modelling approach applied to a number of prototypical alloy systems, with a particular focus on medium- and high-entropy alloys. We also examine the physical implications of a number of underlying assumptions of the modelling approach, including the importance of including the effects of rearrangement of charge due to an applied chemical fluctuation, as well as the effect of the treatment of a system's magnetic state on modelling outcomes. Chapter 8 then examines the $Fe_xGa_{1-x}$ (Galfenol) system as an example of an alloy where atomic arrangements can have a profound impact on predicted magnetic properties. (This is perhaps a little self-indulgent; Galfenol was the first system I studied as a Ph.D. student.) Finally, Chap. 9 summarises the results and provides an outlook on future work.

The physics of metals, alloys, and magnetic systems is a rich and fascinating subject area. Not only are many fundamental physical phenomena beautifully illustrated, but research in this area is often highly relevant for any number of exciting technological applications. It is my hope that my enthusiasm for this subject area carries across over the course of this book and that my (small) contribution to this research area proves useful to others working in the field.

Dr. Christopher D. Woodgate
Department of Physics
University of Warwick
Coventry, UK

# Acknowledgements

Thanks must above all else go to my parents, brother, and wider family for putting up with me on the long and winding journey towards the production of this book. They have, without fail, been caring, helpful, supportive, and (above all) tolerant. I could not have done it without them.

Thanks must also go Prof. Julie B. Staunton, who supervised me, first as a Ph.D. student, then as a postdoctoral researcher, for the time during which I conducted this work. She has been patient, persistent, and enthusiastic at all stages. Julie has, I daresay, probably forgotten more about the electronic structure of metals and alloys than I will ever know. I am indebted to her for the time and effort she has put in to see this project through to its completion.

I am also extremely grateful to Dr. Zachary Evenson, Senior Editor at Springer Nature, who gave me the opportunity to write this book and has been immensely supportive through all stages of its production. Zach has a keen scientific mind, an excellent eye for detail, and is a superb communicator. It has been a pleasure to work with him on this project.

Finally, thanks are due to contemporaries, colleagues, and collaborators, both at Warwick and also further afield. Endless enjoyable, helpful, and probing discussions have deepened my understanding on a range of topics, and I consider myself fractionally less ignorant than when I began my studies. At Warwick, I am particularly grateful to Dr. Nicholas d'Ambrumenil, who supervised my fourth-year project while I was an undergraduate; and to Prof. David Quigley, whose office is just around the corner from mine and who is always willing to 'talk physics' late on a Friday afternoon. Further afield, I would also like to express my gratitude to Prof. Laura H. Lewis of Northeastern University, Boston, who constantly (supportively) probed my understanding of various aspects of my work during the latter portion of my research.

I acknowledge funding from the UK Engineering and Physical Sciences Research Council (EPSRC), which took the form of a Ph.D. studentship within the EPSRC-supported Centre for Doctoral Training in Modelling of Heterogeneous Systems (https://warwick.ac.uk/hetsys), Grant No. EP/S022848/1. Computing facilities were provided by the Scientific Computing Research Technology Platform (SCRTP) of the University of Warwick.

# Contents

# Acronyms

| | |
|---|---|
| ALRO | Atomic long-range order. Atoms arranged in a regular, periodic arrangement |
| ASRO | Atomic short-range order. Short-range correlations between atoms occurring above any atomic ordering temperature |
| bcc | Body-centred-cubic (lattice). A cubic crystal structure with atoms located on the corners and at the centre of a cube. The packing is less efficient than the fcc structure |
| BSF | Bloch spectral function. The Fourier transform of the real-space Green's function of a given electronic system as described within the KKR formalism. In an ordered system, this is equivalent to the usual dispersion relation, *i.e.* the bandstructure. In a disordered system, the bands are 'smeared' and the BSF is viewed as a **k**-resolved DoS function |
| BZ | (First) Brillouin Zone. The primitive unit cell in reciprocal space |
| CPA | Coherent Potential Approximation. A method of constructing the Green's function of an effective medium approximating the electronic properties of a disordered system (typically a substitutional alloy) |
| DFT | Density Functional Theory. A computational, quantum-mechanical modelling method used to investigate the electronic structure of materials, molecules, and other many-body systems |
| DLM | Disordered Local Moment (picture). A DFT-based description of the electronic structure and magnetic properties of magnetic materials above their Curie temperature. Averages over possible spin configurations are evaluated via application of the CPA |
| DoS | Density of States. The number of states per unit energy in a given many-body system |
| fcc | Face-centred-cubic (lattice). A close-packed, cubic crystal structure with atoms located on the corners and face-centres of a cube |
| GGA | Generalised Gradient Approximation. An approximation to the exchange-correlation energy within DFT based on knowledge of the electron density and its spatial derivatives at a given point. Generally |

| | |
|---|---|
| | considered a 'higher' level of theory than the LDA, it is frequently more accurate when determining lattice constants of materials |
| hcp | Hexagonal-close-packed (close-packed crystal structure). A close-packed crystal structure with underlying hexagonal lattice and a two-atom basis |
| HEA | High-entropy alloy. An alloy with four or more chemical species mixed in roughly equal ratios |
| IBZ | Irreducible Brillouin Zone. The wedge of the (first) BZ obtained by reducing the full BZ by all of the symmetries in the point group of the lattice |
| KKR | Korringa–Kohn–Rostoker. A formulation of DFT where the eigenvalue-eigenvector problem is rephrased in terms of electrons propagating through a crystal lattice and scattering from localised, ion-centred potentials. This naturally leads to the expression of key quantities in terms of Green's functions. The KKR formalism proves particularly effective at describing systems with lattice-based disorder |
| KKR-CPA | The CPA as applied in the KKR formulation of DFT |
| LDA | Local Density Approximation. An approximation to the exchange-correlation energy within DFT based solely on knowledge of the electron density at a given point in space |
| LSDA | Local Spin-Density Approximation. A spin-polarised LDA; relevant for magnetic systems |
| MEA | Medium-entropy alloy. An alloy with three chemical species mixed in roughly equal ratios |

# Chapter 1
# Introduction

**Abstract** The motivation for studying the physics of alloy formation is outlined, and the need for computationally efficient modelling approaches discussed. The methodology presented in this work is set in the context of alternative approaches. There is a brief discussion of the layout of the remainder of the book, and also some discussion of alloy naming conventions and nomenclature.

## 1.1 Alloys

Humankind has been alloying metals for millennia, with the overarching aim of producing materials which have favourable properties for applications. Indeed, entire periods of human prehistory are characterised by the use of alloys—first the Bronze Age[1] then the Iron Age[2] [1]. Materials properties which can be improved by alloying include structural properties such as strength, hardness, and ductility, but also chemical properties such as corrosion resistance, and favourable electrical or magnetic properties [2]. Indeed, alloys are ubiquitous in modern society; we eat with stainless steel cutlery, drive cars with 'alloy' wheels, and use solder to join circuits and form electrical connections.

It is not always true, however, that a given combination of elements will alloy 'well'. There are numerous examples of metal pairings which are understood to phase segregate when mixed, *i.e.*, the material spatially decomposes into regions rich (depleted) in one metal and depleted (rich) in the other. In addition, for those systems which *do* alloy, it is necessary to understand whether they form a disordered 'solid solution', or whether instead an 'intermetallic' phase emerges.[3]

---

[1] Bronze is an alloy of copper and tin.

[2] To have superior mechanical properties to bronze, iron must be alloyed at a low concentration with carbon to improve its malleability and ductility.

[3] An intermetallic phase is one in which atoms of different chemical species form an ordered, crystalline, periodic arrangement.



C. D. Woodgate, *Modelling Atomic Arrangements in Multicomponent Alloys*,
Springer Series in Materials Science 346,
https://doi.org/10.1007/978-3-031-62021-8_1

While a large number of alloys are 'binary', *i.e.,* consisting of a mixture of two principal elements, a much larger proportion of the alloy composition space is associated with 'multicomponent' alloys, *i.e.,* those containing three or more principal constituent elements. Indeed, austenitic stainless steel is a three-component alloy typically consisting of approximately 70% Fe, 20% Cr and 10% Ni by weight [3], while additional elements such as C, Mo, and Ti are sometimes added to improve particular properties. Other good examples of 'multicomponent' alloys are the Heusler compounds [4], such as $Cu_2MnAl$, $Fe_2VAl$, and $Co_2NiGa$.

Now, it could be argued that stainless steel is an alloy with only one 'principal element', Fe at a concentration of around 70%, and that Heusler compounds are frequently 'ternary' systems where (typically) a single element represents 50% of the total composition by concentration, making these systems a relatively limited class of materials. However, a recent development in materials science has been the discovery of the so-called medium- and high-entropy alloys (HEAs), first synthesised by Cantor *et al.* [5] and Yeh *et al.* [6] in 2004. These materials, sometimes also referred to as multi-principal element alloys, are systems in which three or more elements are combined in near-equal ratios to form a single-phase solid solution [7–9]. The convention generally taken is that alloys containing three principal elements (sometimes four) are referred to as 'medium-entropy', while those containing four or more (sometimes five or more) elements are referred to as high-entropy. This is because the so-called 'entropy of mixing' for an equiatomic system increases with an increasing number of chemical species. Indeed, Yeh *et al.* suggested that these systems be referred to as 'high-entropy' alloys as they attributed the thermodynamic stability of these compounds to be due to the large entropic contribution to the system's free energy [6]. Numerous examples of these single-phase systems have been discovered, and often they possess exceptional physical properties, such as high fracture-resistance [10], exceptional tensile strength [11], and excellent resistance to corrosion [12], oxidisation [13], and irradiation [14]. In addition, several examples of superconducting high-entropy alloys have been discovered [15, 16], and some are also known to have unusually good soft magnetic properties [17].

A key challenge associated with the design and manufacture of new alloys, and one which is particularly pertinent in the space of multicomponent alloys, is the vast space of potential compositions, which we will illustrate now. Limiting our consideration to equiatomic systems (those where constituent elements are combined in equal ratios), and to the top three rows of the transition metals for a total of 30 elements, the number of possible binary systems is given by

$$\frac{30!}{28!2!} = 435. \tag{1.1}$$

This is already a fairly large number, but it gets much larger if we consider the number of possible equiatomic ternaries,

$$\frac{30!}{27!3!} = 4,060. \tag{1.2}$$

For equiatomic four-component systems the number of possible combinations is 27,405, while for five-component systems it is 142,506. Once we include additional elements not in the main $3d$ block, and once we lift the restriction that we consider only equiatomic systems but instead that we allow atomic ratios to vary, we see that the space of candidate materials is truly vast. While it is true that we do not expect a large number of these candidate systems to form single-phase systems, it is also true that there are undoubtedly untested compositions out there with exceptional physical properties for particular applications. It is therefore desirable to develop improved fundamental physical understanding of the mechanisms driving chemical ordering/segregation in these systems, as well as to develop suitable modelling approaches for exploring this composition space.

In the process of alloy design, there are a number of problems to overcome. First, it is necessary to ascertain whether a suggested composition of metals will form an alloy at all, or whether they will phase segregate. One then needs to obtain a description of the material's underlying crystal structure, and assess whether the system will be chemically disordered or form an ordered intermetallic phase. Physical properties and quantities of interest can then be measured or calculated. Finally, we are often interested in precisely fine-tuning the composition of a given alloy to optimise a particular property. These problems are often laborious to overcome experimentally, and as a result we look to theory and computation to explore the parameter space rapidly and effectively.

In this work, we will outline a first-principles-based[4] approach for modelling atomic arrangements in multicomponent alloys, which is based on a perturbative analysis of the internal energy of the disordered alloy as evaluated via density functional theory (DFT) calculations of materials' electronic structure [18]. (The formalism will, of course, be applicable to binary alloys, too.) In particular, by using the coherent potential approximation (CPA) [19] to construct an effective medium which mimics the electronic structure of the disordered alloy, the approach aims to be computationally efficient, while also capturing key aspects of the physics of alloy formation and the details of their origins in terms of materials' underlying electronic structure. For a given system, the formalism aims to predict whether a given composition will phase segregate, form a single phase solid solution, or form an intermetallic phase following suitable heat treatment. Once this description of the atomic arrangements is obtained, it is then possible to go on to compute physical properties of interest based on these arrangements. In our examples, we will be primarily interested in alloys for magnetic applications, although the framework is much more general and can be applied to systems for a range of applications.

This methodology is perhaps best-viewed as complementary to alternative modelling approaches. In the context of multicomponent alloys, the materials modelling community has developed a range of methodologies for studying these systems [20–22]. Examples of such approaches include semi-empirical schemes such as CALPHAD [23], but also those based on DFT calculations of materials electronic structure. The DFT-based schemes include large-scale 'brute force' studies [24,

[4] *I.e.*, an approach in which we need not be reliant on empirical data to make predictions.

25] but also development of a range of interatomic potentials [26], including cluster expansions [27] and machine-learned interatomic potentials [28–30]. There are also a number of approaches based on effective medium theories such as the CPA [31–39], although frequently these alternative CPA-based methodologies handle effects such as charge rearrangement between elements in some approximate manner. The methodology presented in this work, while still CPA-based, includes fully the effects of charge-rearrangement and charge response and is therefore able to handle a wider range of alloy systems with comparative ease. Ongoing work also demonstrates that this computationally efficient, lattice-based modelling approach can also be used to inform alternative modelling approaches, for example by providing physically motivated atomic arrangements for use in training datasets for machine-learned interatomic potentials [40], improving the efficiency of the training process.

The remainder of this work is structured as follows. In Chap. 2, we outline how we will apply the techniques of statistical physics to analyse atomic arrangements in alloys in a multicomponent setting. This chapter leaves the internal energy of the alloy *unspecified*, and no background in electronic structure theory is assumed, with the aim of making it accessible to those with a non-specialist background. We will highlight the importance of the Onsager cavity field, which extends the temperature range for which the theory is valid. Some illustrative examples are given for an alloy with an internal energy given by a simple, pairwise model, and atomistic modelling techniques for use with this model are outlined.

In Chap. 3, we then build up the relevant tools of electronic structure to lead to a first-principles-based description of atomic arrangements in multicomponent alloys. We start by giving a brief overview of the theory of electronic structure focussing on DFT [18] and, in particular, the Korringa-Kohn-Rostoker (KKR) [19] formulation thereof. This is to outline how it is possible to describe the electronic 'glue' driving atomic ordering and producing magnetic phenomena within a first-principles framework, *i.e.*, without empirically-determined parameters. We then outline how the CPA is used to model the electronic structure of disordered solid solutions [41, 42]. We will also outline the key details of the disordered local moment (DLM) picture [43–45], which enables the paramagnetic state of an alloy to be described within the CPA. The paramagnetic state is important because, typically, multicomponent alloys are processed at temperatures well above their Curie temperature, the temperature at which they experience a magnetic disorder-order transition. Having developed the relevant background, we will then outline the $S^{(2)}$ theory of multicomponent alloys as given by Khan, Staunton, and Stocks [46], and state the set of coupled equations which must be solved self-consistently to compute the response of the CPA medium to an applied chemical perturbation. We will also outline the key features of the so-called 'band-only' approximation, in which there is no charge-transfer or charge-response due to an imposed chemical fluctuation, which simplifies the structure of the equations greatly, but in many cases can be shown to yield unphysical results. Finally, at the end of this chapter, the electronic structure of some prototypical metals and alloys will be discussed to give insight.

Moving onto applications of the methodology, Chap. 4 describes results applying the modelling approach to the refractory high-entropy alloys, a prototypical family of multicomponent alloys forming on the bcc lattice in which there are no magnetic elements [47]. It is shown that the approach yields results which agree well with existing literature and correctly predicts the phase behaviour of these complex systems. We also give insight into the physical origins of atomic ordering in terms of the materials' electronic structure.

In a development relevant to materials for applications, in Chap. 5, we then consider the addition of Ti, at a low concentration, to the aforementioned refractory systems [48]. Ti is a suggested additive to these systems to improve their room-temperature ductility while maintaining favourable high-temperature properties. However, during processing, Ti-containing systems develop chemical inhomogeneities. In good agreement with experiment, our approach predicts competition between chemical ordering and chemical segregation in these systems, leading to complex, multiphase behaviour. As for the refractories of the previous chapter, these predictions are set in context of the underlying electronic structure of the disordered solid solutions, enabling physical intuition to be drawn.

Proceeding, Chap. 6 describes results for the Cantor-Wu medium- and high-entropy alloys, another well-known family of multicomponent alloys, forming on the fcc lattice and containing magnetic elements such as Fe, Ni, and Co [49]. In this chapter the paramagnetic state is modelled within the disordered local moment (DLM) picture, the picture which we believe is most physical and reflective of materials processing conditions for these systems. Once again, the results obtained are compared to existing literature, both theoretical and experimental, and good agreement is obtained.

To illustrate the importance of correct treatment of the magnetic state of a material when modelling atomic arrangements, in Chap. 7 we compare results of the phase stability analysis of of the Cantor-Wu systems modelled assuming a paramagnetic state with the same analysis performed assuming the systems are in either a non-magnetic or magnetically ordered (ferrimagnetic) state [50]. It is shown that dramatically different phase behaviour is predicted depending on how the magnetic state of a material is modelled. These results have implications both for materials modelling, by emphasising the importance of the correct treatment of a material's magnetic state during simulation, but also for materials processing, by suggesting that it may be possible to tune atomic arrangements and materials properties in some multicomponent alloy compositions by annealing samples in an applied magnetic field.

To demonstrate how atomic arrangements in magnetic alloys can affect magnetic properties, in Chap. 8, an holistic approach for modelling the phase behaviour and subsequent Curie temperature and magnetoelastic constants of $Fe_{1-x}Ga_x$ (Galfenol) is given [51]. It is shown that atomic ordering can alter not only the Curie temperature of the system, but also its predicted magnetoelastic constants, which quantify how magnetostrictive it is.[5]

[5] Magnetostriction is a phenomenon where the dimensions of shape of a material change during magnetisation.

Finally, in Chap. 9, the results are summarised, their implications examined, and an outlook on further work is given.

## 1.2 A Note on Nomenclature: Multicomponent, Multi-Principal-Element, or High-Entropy?

The alloy literature is awash with competing terms and definitions to describe multicomponent alloys. Terms such as 'multicomponent', 'multi-principal-element', 'complex-concentrated' and 'high-entropy' are used by different authors, often to mean slightly different things. In this work, we will use the term *multicomponent* to describe an alloy with three or more alloying elements. We consider the medium- and high-entropy alloys a subset of the space of multicomponent alloys. We consider an alloy composed of three elements at near-equal ratios to be a 'medium-entropy' alloy, while an alloy consisting of four or more elements at near-equal ratios is considered to be 'high-entropy'.

## 1.3 Naming Convention: In Which Order Should Elements be Listed?

Even more confusing than the abundance of terms describing multicomponent alloys is the near-complete absence of a naming convention for multicomponent systems. It is possible to find examples in the literature of the same alloy being referred to by four or five different names. In this work, when discussing a medium- or high-entropy alloy, alloying elements will be listed in order of increasing atomic number, from smallest to largest. So, for example, the three-component alloy consisting of Cr, Co, and Ni in equal ratios is referred to as CrCoNi, while the five-component alloy formed from V, Nb, Mo, Ta, and W is referred to as VNbMoTaW.

## References

1. L.E. Murr, in *Handbook of Materials Structures, Properties, Processing and Performance* (Springer International Publishing, Cham, 2015), pp. 3–9. https://doi.org/10.1007/978-3-319-01815-7
2. W. Pfeiler (ed.), *Alloy Physics: A Comprehensive Reference*, 1st edn. (Wiley, 2007). https://doi.org/10.1002/9783527614196
3. P. Marshall, *Austenitic Stainless Steels: Microstructure and Mechanical Properties* (Elsevier Applied Science Publ, London, 1984)
4. C. Felser, A. Hirohata (eds.), *Heusler Alloys: Properties, Growth, Applications, Springer Series in Materials Science*, vol. 222 (Springer International Publishing, Cham, 2016). https://doi.org/10.1007/978-3-319-21449-8

5. B. Cantor, I.T.H. Chang, P. Knight, A.J.B. Vincent, Mater. Sci. Eng. A **375–377**, 213 (2004). https://doi.org/10.1016/j.msea.2003.10.257
6. J.W. Yeh, S.K. Chen, S.J. Lin, J.Y. Gan, T.S. Chin, T.T. Shun, C.H. Tsau, S.Y. Chang, Adv. Eng. Mater. **6**(5), 299 (2004)
7. D.B. Miracle, O.N. Senkov, Acta Materialia **122**, 448 (2017). https://doi.org/10.1016/j.actamat.2016.08.081
8. M.C. Gao, J.W. Yeh, P.K. Liaw, Y. Zhang (eds.), *High-Entropy Alloys* (Springer International Publishing, Cham, 2016). https://doi.org/10.1007/978-3-319-27013-5
9. E.P. George, D. Raabe, R.O. Ritchie, Nat. Rev. Mater. **4**(8), 515 (2019)
10. W. Li, P.K. Liaw, Y. Gao, Intermetallics **99**, 69 (2018). https://doi.org/10.1016/j.intermet.2018.05.013
11. Z. Fu, L. Jiang, J.L. Wardini, B.E. MacDonald, H. Wen, W. Xiong, D. Zhang, Y. Zhou, T.J. Rupert, W. Chen, E.J. Lavernia, Sci. Adv. **4**(10), eaat8712 (2018). https://doi.org/10.1126/sciadv.aat8712
12. Y. Qiu, S. Thomas, M.A. Gibson, H.L. Fraser, N. Birbilis, Npj Mater. Degrad. **1**(1), 15 (2017). https://doi.org/10.1038/s41529-017-0009-y
13. B.R. Anne, S. Shaik, M. Tanaka, A. Basu, S.N. Appl, Sci. **3**(3), 366 (2021)
14. O. El-Atwani, N. Li, M. Li, A. Devaraj, J.K.S. Baldwin, M.M. Schneider, D. Sobieraj, J.S. Wróbel, D. Nguyen-Manh, S.A. Maloy, E. Martinez, Sci. Adv. **5**(3), eaav2002 (2019). https://doi.org/10.1126/sciadv.aav2002
15. P. Koželj, S. Vrtnik, A. Jelen, S. Jazbec, Z. Jagličić, S. Maiti, M. Feuerbacher, W. Steurer, J. Dolinšek, Phys. Rev. Lett. **113**(10), 107001 (2014)
16. L. Sun, R.J. Cava, Phys. Rev. Mater. **3**(9), 090301 (2019)
17. Y. Zhang, T. Zuo, Y. Cheng, P.K. Liaw, Sci. Rep. **3**(1), 1455 (2013)
18. R.M. Martin, *Electronic Structure: Basic Theory and Practical Methods* (Cambridge University Press, Cambridge, UK, 2004)
19. J.S. Faulkner, G.M. Stocks, Y. Wang, *Multiple Scattering Theory: Electronic Structure of Solids*, 1st edn. (IOP Publishing, Bristol, UK, 2018). https://doi.org/10.1088/2053-2563/aae7d8
20. M. Widom, J. Mater. Res. **33**(19), 2881 (2018)
21. A. Ferrari, B. Dutta, K. Gubaev, Y. Ikeda, P. Srinivasan, B. Grabowski, F. Körmann, J. Appl. Phys. **128**(15), 150901 (2020)
22. A. Ferrari, F. Körmann, M. Asta, J. Neugebauer, Nat. Comput. Sci. **3**(3), 221 (2023)
23. C. Zhang, Y. Yang, MRS Bull. **47**(2), 158 (2022)
24. A. Tamm, A. Aabloo, M. Klintenberg, M. Stocks, A. Caro, Acta Materialia **99**, 307 (2015). https://doi.org/10.1016/j.actamat.2015.08.015
25. J. Ding, Q. Yu, M. Asta, R.O. Ritchie, Proceed. Nat. Acad. Sci. **115**(36), 8919 (2018)
26. A.D. Kim, M. Widom, Phys. Rev. Mater. **7**(6), 063803 (2023)
27. A. Fernández-Caballero, J.S. Wróbel, P.M. Mummery, D. Nguyen-Manh, J. Phase Equil. Diff. **38**(4), 391 (2017)
28. T. Kostiuchenko, F. Körmann, J. Neugebauer, A. Shapeev, Npj Comput. Mater. **5**(1), 55 (2019). https://doi.org/10.1038/s41524-019-0195-y
29. C.W. Rosenbrock, K. Gubaev, A.V. Shapeev, L.B. Pártay, N. Bernstein, G. Csányi, G.L.W. Hart, Npj Comput. Mater. **7**(1), 24 (2021). https://doi.org/10.1038/s41524-020-00477-2
30. S. Ghosh, V. Sotskov, A.V. Shapeev, J. Neugebauer, F. Körmann, Phys. Rev. Mater. **6**(11), 113804 (2022)
31. A.V. Ruban, S. Shallcross, S.I. Simak, H.L. Skriver, Phys. Rev. B **70**(12), 125115 (2004)
32. P. Singh, A.V. Smirnov, D.D. Johnson, Phys. Rev. B **91**(22), 224204 (2015)
33. F. Körmann, A.V. Ruban, M.H. Sluiter, Mater. Res. Lett. **5**(1), 35 (2017). https://doi.org/10.1080/21663831.2016.1198837
34. P. Singh, A.V. Smirnov, D.D. Johnson, Phys. Rev. Mater. **2**(5), 055004 (2018)
35. B. Schönfeld, C.R. Sax, J. Zemp, M. Engelke, P. Boesecke, T. Kresse, T. Boll, T. Al-Kassab, O.E. Peil, A.V. Ruban, Phys. Rev. B **99**(1), 014206 (2019)
36. T. Kostiuchenko, A.V. Ruban, J. Neugebauer, A. Shapeev, F. Körmann, Phys. Rev. Mater. **4**(11), 113802 (2020)

37. P. Singh, A.V. Smirnov, A. Alam, D.D. Johnson, Acta Materialia **189**, 248 (2020). https://doi.org/10.1016/j.actamat.2020.02.063. https://www.sciencedirect.com/science/article/pii/S1359645420301774
38. F. Körmann, T. Kostiuchenko, A. Shapeev, J. Neugebauer, Phys. Rev. Mater. **5**(5), 053803 (2021)
39. Y. Zhou, P. Srinivasan, F. Körmann, B. Grabowski, R. Smith, P. Goddard, A.I. Duff, Phys. Rev. B **105**(21), 214302 (2022)
40. L. Shenoy, C.D. Woodgate, J.B. Staunton, A.P. Bartók, C.S. Becquart, C. Domain, J.R. Kermode, Phys. Rev. Mater. **8**(3), 033804 (2024)
41. P. Soven, Phys. Rev. **156**(3), 809 (1967)
42. D.D. Johnson, D.M. Nicholson, F.J. Pinski, B.L. Györffy, G.M. Stocks, Phys. Rev. B **41**(14), 9701 (1990)
43. A.J. Pindor, J. Staunton, G.M. Stocks, H. Winter, J. Phys. F: Metal Phys. **13**(5), 979 (1983)
44. J. Staunton, B. Gyorffy, A. Pindor, G. Stocks, H. Winter, J. Magn. Magn. Mater. **45**(1), 15 (1984)
45. B.L. Gyorffy, A.J. Pindor, J. Staunton, G.M. Stocks, H. Winter, J. Phys. F: Metal Phys. **15**(6), 1337 (1985)
46. S.N. Khan, J.B. Staunton, G.M. Stocks, Phys. Rev. B **93**(5), 054206 (2016)
47. C.D. Woodgate, J.B. Staunton, Phys. Rev. Mater. **7**(1), 013801 (2023)
48. C.D. Woodgate, J.B. Staunton, J. Appl. Phys. **135**(13), 135106 (2024)
49. C.D. Woodgate, J.B. Staunton, Phys. Rev. B **105**(11), 115124 (2022)
50. C.D. Woodgate, D. Hedlund, L.H. Lewis, J.B. Staunton, Phys. Rev. Mater. **7**(5), 053801 (2023)
51. G.A. Marchant, C.D. Woodgate, C.E. Patrick, J.B. Staunton, Phys. Rev. B **103**(9), 094414 (2021)

# Chapter 2
# Statistical Physics of Multicomponent Alloys

**Abstract** We consider the application of techniques from statistical physics to problems concerning the phase behaviour of multicomponent alloys. We begin by introducing the *site-wise occupancies* (discrete) and *site-wise concentrations* (continuous), which are variables used to concisely specify the configuration (or a set of configurations) of atoms in an alloy. The concept of a free energy is then briefly outlined, and a discussion of Landau theory is given. The majority of the chapter is then dedicated to the derivation of a tractable form of the alloy free energy for use in a Landau-type linear response theory, and to the physical interpretation of results obtained using this theory. The chapter concludes by presenting some intuitive examples of 'toy' alloy systems where the internal energy takes a simple, pairwise form, and analysing these using the Landau theory in conjunction with lattice-based Monte Carlo simulations.

## 2.1 Introductory Remarks

We are concerned with the physics of *alloys*. For our purposes, an alloy is crystalline material composed of two or more different metallic elements, which we will refer to as chemical species, or 'components'. Although the underlying structure is crystalline, no restrictions are placed on which species occupy which lattice site, or on how many (or few) species are considered.

In this work, only substitutional alloys are considered. A substitutional alloy is a crystalline system with an underlying lattice (and basis) which, in principle, extends infinitely in all spatial directions. (Of course, for practical calculations this is replaced with a finite, periodic system.) All lattice sites are occupied by atoms, and atoms may only sit on lattice sites. We may 'substitute' an atom on one site for one of a different species, but we may not consider introducing an atom sitting away from a lattice site, and we may not consider removing an atom from a site and not replacing it with another. This book deals mainly with alloys with a monatomic basis whose underlying crystal structure is close-packed, *e.g.*, fcc, bcc, or (potentially) hcp.

It is worth noting that lots of these restrictions can be lifted. In so-called interstitial alloys, atoms can sit off metal lattice sites and effectively squeeze into the 'gaps'



C. D. Woodgate, *Modelling Atomic Arrangements in Multicomponent Alloys*, Springer Series in Materials Science 346,
https://doi.org/10.1007/978-3-031-62021-8_2

between them. A good example of this is the addition of carbon to steel. The carbon atoms are sufficiently small to sit in the 'gaps' between lattice sites. Assuming you have an understanding of where the interstitial atoms sit relative to the on-lattice atoms, it is possible to consider these systems within our formalism, but we will not consider such systems here. It is also perfectly reasonable to consider alloys with vacant lattice sites, referred to as vacancies. Again, these can be described within our formalism by describing vacancies as a chemical species at a (very low) concentration. However, in both cases (interstitials and vacancies) no account of local lattice relaxations will be given. This is a limitation of the formalism but, as we shall see, a large number of alloy systems are described successfully. In addition, the restriction to on-lattice modelling helps keep the computational cost of the approach low, an important consideration at a time when the environmental cost of high-performance computing is under increasing scrutiny [1]. It should also be said that ongoing research, not included in this work, suggests that we can study interstitials and vacancies using our approach and obtain reasonably physical results [2].

When discussing atomic arrangements in alloys, the state of matter to have in mind is the so-called 'solid solution'. This is a state below the melting temperature of the alloy in which atoms sit on lattice sites but are (relatively) free to move from one site to another by swapping with each other.[1] When an alloy is annealed, it is held at a temperature lower than its melting temperature but high enough that atomic swaps in the solid solution occur with relative ease to enable ordered structures to form. When an ordered structure like this forms, it is generally referred to as an 'intermetallic' compound. Sometimes, an alloy is rapidly quenched from a high temperature to 'freeze' the high temperature structure in place. It is also common to cool the material slowly to maximise atomic ordering. Further, a number of modern processing techniques impose particular physical conditions, *e.g.*, mechanical strain or a magnetic field during the annealing process, to promote the desired atomic ordering.

The pertinent questions to ask about atomic arrangements in alloys are three-fold. The first question is: what is the equilibrium state of the material at a given temperature? That is, assuming I anneal an alloy at a given temperature and then quench it to room temperature (where atoms are essentially frozen in place) how will my atoms be arranged? What is the underlying crystal structure, how are the atoms of different species arranged on that structure, and is there some kind of ordering occurring between atoms of different species?

The second question is: how long does it take for my atoms to arrange themselves into the determined equilibrium structure? This is a kinetics problem, and is all about assessing the height of the energetic barrier that has to be climbed for atoms to 'jiggle' around and swap with their neighbours. This is highly relevant because, no matter what the predicted equilibrium configuration, to be experimentally realisable the predicted atomic arrangement has to be accessible on a human timescale. This is broadly a question of temperature; at high temperatures atomic swaps occur with

[1] We will not concern ourselves with exactly *how* these swaps occur here. Elemental diffusion in metals and alloys is a complex phenomenon and is covered in depth elsewhere, *e.g.*, in [3].

relative ease, while at lower temperatures the lack of thermal energy in the material (and the corresponding lack of atomic 'jiggling') effectively freezes atoms in place.

The third and final question is then: what effect does the arrangement of atoms have on measurable physical properties? Atomic arrangements can affect a wide range of materials properties, including electrical properties such as resistivity, mechanical properties such as ductility and hardness, and magnetic properties such as a system's Curie temperature [4]. We would ideally like to be able to describe both the nature of atomic arrangements in a material and the subsequent impact of these arrangements on physical properties within some holistic modelling framework.

In this work, we will seek to answer the first and third questions; what is the equilibrium arrangement of the atoms, and how does this affect physical properties of interest? The second question, about kinetics, we will leave as a question related to temperature; if we predict that atoms will arrange themselves in a certain way at a given temperature, and that temperature is high enough, then we assume that this ordered arrangement will be kinetically accessible.

The aim of this chapter is to apply techniques from statistical physics to the problem of describing the equilibrium arrangement of atoms in an alloy. Inevitably this leads us to evaluating some approximation of the internal energy of the alloy. Eventually, this internal energy will come from first-principles calculations, namely those based on density functional theory (DFT) [5]. However, for the purposes of this chapter, we will leave the form of the internal energy largely unspecified, except for the purposes of giving some intuitive examples to assist in understanding what comes later.

Before diving into the formalism, it might be useful to try to get a handle on the scale of the problem which faces us, and why the tools of statistical physics are appropriate. Let us imagine a model of an alloy on the face-centred cubic lattice, with four cubic unit cells in each coordinate direction, giving a total of 256 atoms, and periodic boundary conditions applied in all three coordinate directions. This size of system is fairly routine for a study employing DFT calculations to evaluate energies as, although not computationally 'cheap', it is easily tractable on a modern computing cluster. Evaluation of the partition function, the key quantity of statistical physics, involves a sum over all possible microstates (*i.e.*, configurations) of the system. The question is: how many possible configurations are there for this system?

Well, assuming we are modelling an equiatomic, binary alloy, the total number of unique on-lattice arrangements of 256 atoms with 128 of type (or species) $A$ and 128 of species $B$ is given by

$$\frac{256!}{(128!)(128!)} \approx 5.77 \times 10^{75}.$$

This is a *big* number. In fact, it's not too far off the Eddington number—an estimate of the number of protons in the observable universe—which was originally calculated to be about $1.57 \times 10^{79}$ [6]. Of course, due to the various symmetries of the system, we expect a vast number of these possible configurations to be degenerate, but at least we have an idea of the problem we are facing.

The situation gets significantly worse if we consider alloys with multiple principle elements. Let us now consider an equiatomic, four component alloy with 64 atoms each of species $A$, $B$, $C$, and $D$. (Some examples of systems such as this will, in fact, be considered later.) The total number of arrangements of our 256 on-lattice atoms is then given by

$$\frac{256!}{(64!)(64!)(64!)(64!)} \approx 3.31 \times 10^{150}.$$

Again, the same arguments about symmetry reducing the number of non-degenerate configurations significantly still apply, but we see how quickly the space of possible configurations can grow once multiple components are considered.

We see, therefore, that the problem of evaluating the partition function directly would be a futile endeavour, and that appropriate approximations must be made. Having motivated the problem and given an illustration of its scale and complexity, let us now begin our exposition in earnest.

## 2.2 Preliminaries

This work keeps broadly to the notation developed by Khan, Staunton and Stocks [7]. Throughout, Greek letters are taken as chemical species labels, and Roman characters as site labels. The number of chemical species (types of atom) in an alloy will be denoted by the letter $s$. The interested reader may also wish to consult the excellent book of Ducastelle [8] for further insight into the statistical physics of alloys; his exposition being much more detailed than the one given here.

Explicitly, we consider a crystal with fixed Bravais lattice $\{\mathbf{R}_i\}$ and associated basis $\{\mathbf{h}_1, \ldots, \mathbf{h}_n\}$. The complete set of site positions is then written $\{\mathbf{R}_i + \mathbf{h}_a\}$. We restrict ourselves to considering case that all site positions are crystallographically equivalent; this will be important in arguments which follow later. The occupancy of a lattice site with composite index $\mathrm{i} = (i, a)$ is then written

$$\xi_{\mathrm{i}\alpha} := \begin{cases} 1, & \text{if site } \mathrm{i} = (i, a) \text{ is occupied by an atom of species } \alpha \\ 0, & \text{if site } \mathrm{i} = (i, a) \text{is } \textit{not} \text{ occupied by an atom of species } \alpha. \end{cases} \tag{2.1}$$

These numbers will be referred to as 'site-wise occupations', or 'site occupation numbers'. The complete set of these numbers, $\{\xi_{\mathrm{i}\alpha}\}$, uniquely specifies a configuration of the system. The overall concentration of each chemical species is then written

$$c_\alpha := \frac{1}{N} \sum_{\mathrm{i}} \xi_{\mathrm{i}\alpha}, \tag{2.2}$$

where $N$ is the total number of lattice sites.

We restrict the system to having one (and only one) chemical species occupying each site, which can be expressed as:

$$\sum_{\alpha} \xi_{\mathrm{i}\alpha} = 1 \tag{2.3}$$

for every lattice site $\mathrm{i} = (i, a)$. It can be helpful to think of each site as having a vector of associated occupation numbers,

$$\xi_{\mathrm{i}\alpha} = (\xi_{\mathrm{i}1}, \ldots, \xi_{\mathrm{i}s}), \tag{2.4}$$

where these numbers are all zero apart from one, indicating the chemical species sitting on that lattice site. Given an $s$-species alloy, the condition that each lattice site is occupied (2.3) reduces us to $(s-1)$ independent occupation numbers. That is,

$$\begin{aligned} \xi_{\mathrm{i}s} &= 1 - \sum_{\alpha=1}^{s-1} \xi_{\mathrm{i}\alpha} \\ &= 1 - \sum_{\alpha}{}' \xi_{\mathrm{i}\alpha}, \end{aligned} \tag{2.5}$$

where the prime denotes a sum over species indices excluding $s$. Species $s$ is then referred to as the 'host'. The choice of host is arbitrary and any physically meaningful results must not depend on this choice. It is also worth observing that this makes binary alloys (those with two component species) particularly simple; at each lattice site there is only one independent occupation number.

Given an ensemble of configurations, the natural quantity of interest describing how atoms are arranged is the ensemble average of the site occupancy,

$$c_{\mathrm{i}\alpha} := \langle \xi_{\mathrm{i}\alpha} \rangle, \tag{2.6}$$

which is an atomic long-range order parameter. We will refer to these quantities as the 'site-wise concentrations'. Note that, by 2.3, we have that

$$\sum_{\alpha} c_{\mathrm{i}\alpha} = \sum_{\alpha} \langle \xi_{\mathrm{i}\alpha} \rangle \tag{2.7}$$

$$= \langle \sum_{\alpha} \xi_{\mathrm{i}\alpha} \rangle \tag{2.8}$$

$$= 1, \tag{2.9}$$

so it is possible to write

$$c_{\mathrm{i}s} = 1 - \sum_{\alpha}{}' c_{\mathrm{i}\alpha}, \tag{2.10}$$

in the same spirit as for 2.5. And, again, in a binary alloy, there is only one independent site-wise concentration variable.

In the high-temperature limit in which the alloy is disordered, the site-wise concentrations are spatially homogeneous, *i.e.*,

$$c_{i\alpha} \equiv c_\alpha, \tag{2.11}$$

where we recall that $c_\alpha$ is the total overall concentration of each species, specified in 2.2. However, in general we expect that at some finite temperature $T_{us}$, where 'us' is shorthand for 'unstable,' this symmetry will be broken and the site-wise concentrations will acquire a spatial dependence. It is often convenient to express the spatially dependent site-wise concentrations as a fluctuation about the homogeneous reference state,

$$c_{i\alpha} = c_\alpha + \Delta c_{i\alpha}, \tag{2.12}$$

where $\Delta c_{i\alpha}$ is the fluctuation. Moreover, given the translational symmetry of the underlying lattice, it is natural to express these quantities in reciprocal space, as pioneered by Khachaturyan [9, 10]. The site-wise concentrations are then written as

$$c_{i\alpha} = c_\alpha + \sum_{\mathbf{k}} e^{i\mathbf{k}\cdot\mathbf{R}_i} \Delta c_{a\alpha}(\mathbf{k}), \tag{2.13}$$

where the sum is taken over $\mathbf{k}$ in the first Brillouin zone, and the $\Delta c_{a\alpha}(\mathbf{k})$ are thought of as 'concentration waves'. These concentration waves can be used to describe ordered structures in a clear and concise fashion, and it is possible to classify a huge number of potential chemical orderings in this way.

As examples, we consider two prototypical alloy systems, $Cu_xZn_{1-x}$ and $Cu_x Au_{1-x}$. The Cu-Zn system is more commonly referred to as brass and, in the range $x \approx 0.5$ forms on a body centred cubic (bcc) lattice, which has the *Strukturbericht* designation[2] A2. This is not the most energetically favourable configuration for the atoms, however, as they prefer to form a CsCl structure, *Strukturbericht* designation B2, which partitions the bcc structure into two interpenetrating simple cubic sublattices, one of which will be occupied by Zn atoms, and one of which will be occupied by Cu atoms. The A2 and B2 structures are visualised in Fig. 2.1. This ordered structure is also referred to as $\beta$-brass. The bcc system has a monatomic basis so we drop Roman index $a$. The overall concentrations are written $c_\alpha = (1/2, 1/2)$. We can describe the B2 ordering by an order parameter $\eta$ and a concentration wave with

---

[2] The *Strukturbericht* designations are a crystallographic classification scheme widely used in the alloy literature. They offer a convenient shorthand for representing a variety of disordered and ordered structures. A comprehensive encyclopedia of crystallographic prototypes is offered by *AFLOW* [11–13].

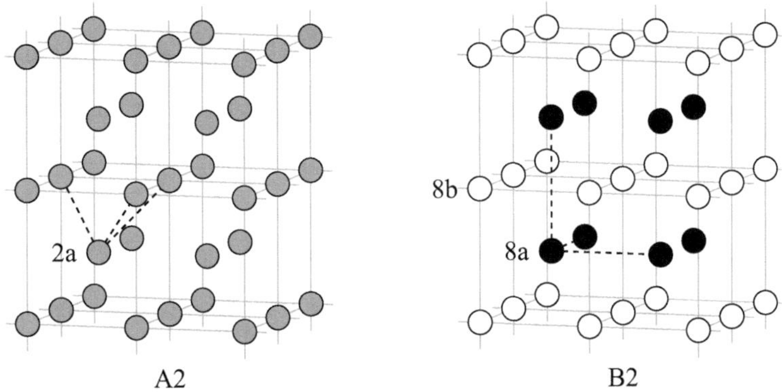


**Fig. 2.1** Visualisation of the A2 (disordered bcc) and B2 (CuZn) structures. The B2 structure is described by modes $\mathbf{k} = (0, 0, 1)$ and equivalent, and $\Delta c_\alpha = \frac{1}{\sqrt{2}}(1, -1)$. Sites are given their Wyckoff labels, and primitive lattice vectors are visualised by dashed lines

$\mathbf{k} = (0, 0, 1)$ and equivalent,[3] and concentration polarisation $\Delta c_\alpha = \frac{1}{\sqrt{2}}(1, -1)$. The lattice site occupancies are then given by

$$c_{i\alpha} = c_\alpha + \eta \sum_{\mathbf{k}} e^{i\mathbf{k}\cdot\mathbf{R}_i} \Delta c_\alpha(\mathbf{k}), \tag{2.14}$$

where $\eta = 0$ corresponds to the disordered, A2, structure, and $\eta = \frac{\sqrt{2}}{12}$ describes the perfect B2 structure. (Note that, although we could appropriately scale $\Delta c_\alpha$ to enforce $0 \leq \eta \leq 1$, we take the convention of normalising the $\Delta c_\alpha$ vector throughout this work.)

The Cu-Au system is an example alloy system forming on the face-centred cubic (fcc) lattice. Here we consider two compositions, $x = 0.75$, *i.e.*, $Cu_3Au$, and $x = 0.5$, *i.e.*, CuAu. The relevant ordered structures are visualised in Fig. 2.2.

For $Cu_3Au$, the energetically favoured configuration of atoms (conventionally referred to as the $Cu_3Au$ structure) has the *Strukturbericht* designation $L1_2$, and consists of Au atoms sitting on the corners of the fcc lattice, with Cu atoms in the face centres. The overall concentrations are $c_\alpha = (3/4, 1/4)$, and the ordered structure is described by a concentration wave with $\mathbf{k} = (0, 0, 1)$ and equivalent, and concentration polarisation $\Delta c_\alpha = \frac{1}{\sqrt{2}}(1, -1)$

Now there is a subtlety; the CuAu system ($x = 0.5$) forms an arrangement of atoms with *Strukturbericht* designation $L1_0$, however the concentration wave describing this structure is still described by $\Delta c_\alpha = \frac{1}{\sqrt{2}}(1, -1)$. This time, though, rather than having $\mathbf{k} = (0, 0, 1)$ *and equivalent*, the system instead is forced by its composition

[3] 'Equivalent' wavevectors are all those wavevectors $\mathbf{k}'$ that can be mapped back to $\mathbf{k}$ by translating by a reciprocal lattice vector, $\mathbf{G}$, *i.e.*, $\mathbf{k} = \mathbf{k}' + \mathbf{G}$ for some reciprocal lattice vector $\mathbf{G}$. Sometimes the set of all wavevectors equivalent to a wavevector $\mathbf{k}_0$ is referred to as the 'star' of $\mathbf{k}_0$.

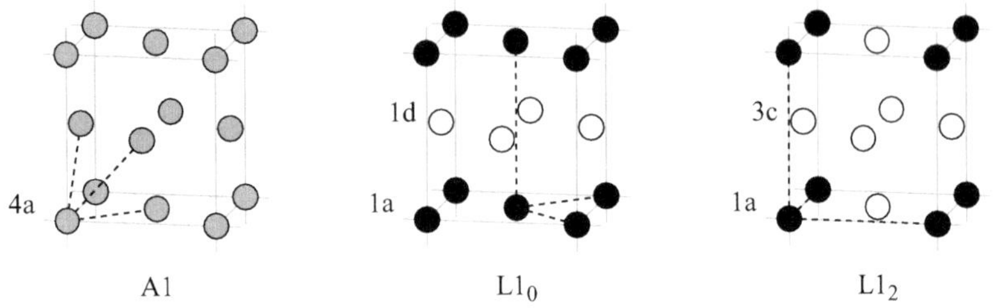


**Fig. 2.2** Visualisation of the A1 (disordered fcc), $L1_0$ (CuAu), and $L1_2$ ($Cu_3Au$) structures. Sites are given their Wyckoff labels, and primitive lattice vectors are visualised by dashed lines

to *pick* a direction (chosen arbitrarily to be the $\hat{\mathbf{z}}$ direction in this example) and form a layered structure along that direction only. Taking $\{\mathbf{k}\} = \{(0, 0, 1), (0, 0, -1)\}$ yields the correct structure.

Finally, we note that this formalism is not only capably of describing the formation of ordered structures, but also of describing phase segregation. An equiatomic binary alloy which phase segregates is described by a concentration wave with $\mathbf{k} = (0, 0, 0)$, and concentration polarisation $\Delta c_\alpha = \frac{1}{\sqrt{2}}(1, -1)$. This represents a concentration wave with (effectively) infinite length, which naturally describes the case of phase segregation.

The versatility of this formalism is clear to be seen, and it remains to use this language and the tools of statistical physics to analyse alloy phase behaviour. Before we proceed in earnest, however, it is necessary to make a small justification. For the purposes of constructing a Landau-type theory of atomic ordering, we need to be able to set $c_{i\alpha}$ arbitrarily and vary them smoothly to perform a Taylor expansion of the alloy free energy. The argument is as follows. We imagine an ensemble of configurations and their associated probabilities, where we *choose* the site-wise concentrations $c_{i\alpha}$ and overall concentration of each species $c_\alpha$. The probability distribution for this ensemble is not necessarily physical and is acquired by minimising the Helmholtz free energy,

$$F = \langle U \rangle - T \langle S \rangle, \tag{2.15}$$

subject to the imposed constraints on the site-wise concentrations. (Of course, by relaxing the imposed site-wise concentrations and only imposing a constraint on the total overall concentration of each species, enforced by the chemical potentials, the physical Boltzmann distribution is recovered.) We see in this way that $c_{i\alpha}$ can be used to define a set of variables which can be continuously varied while preserving a known set of overall concentrations.

### 2.2.1 Host-Dependent and Host-Independent Settings

In what follows, we will be interested in quantities which are function(al)s[4] of the site-wise concentrations, *i.e.*, some $f$ for which $f = f[\{c_{\mathrm{i}\alpha}\}]$. Moreover, we will be interested in evaluating derivatives of these quantities with respect to the site-wise concentrations. The most natural setting for these quantities is a setting where we consider the set of $s$ variables $(c_{\mathrm{i}1}, \ldots, c_{\mathrm{i}s})$, which will be referred to as the 'host-independent' setting. However, for the purposes of performing a Landau-type expansion of the free energy of the system and subsequent analysis, it will often be more convenient to work with the set of variables $(c_{\mathrm{i}1}, \ldots, c_{\mathrm{i}(s-1)})$, which we will refer to as the 'host-dependent' setting.[5] We will switch freely between these two settings as is appropriate to the analysis.

For ease of reading, henceforth we will write any quantity which is to be viewed in the host-independent setting with a 'tilde' on top, $\tilde{f}$, while any quantity which is to be viewed in the host-dependent setting will have the tilde omitted, $f$. Sums over chemical species will be written with a prime, $\sum'$, when the host species is to be omitted, while sums over all chemical species, including the host, will be written without a prime, $\sum$. Quantities which are independent of the choice of setting, such as free energies, will have a tilde omitted always. Appropriate mappings for important quantities between the two setting will be defined presently. Having clarified this important point, we now move on to outline a means by which Landau theory can be applied to multicomponent alloys.

## 2.3 Landau Theory

Given a free energy $F$, Landau theory seeks to write a model free energy as a series expansion around some high-symmetry, high-temperature reference state in terms of a suitable set of order parameters which characterise a phase transition. As has already been outlined, the choice of order parameters here are the site-wise concentrations $\{c_{\mathrm{i}\alpha} = c_\alpha + \Delta c_{\mathrm{i}\alpha}\}$, and $F$ is considered a function(al) of these, *i.e.*, $F = F[\{c_{\mathrm{i}\alpha}\}]$. Performing a Taylor expansion about the homogeneous reference state gives

$$\begin{aligned} F\left[\{c_{\mathrm{i}\alpha}\}\right] =& F\left[\{c_\alpha\}\right] + \sum_{i\alpha} \left.\frac{\partial F}{\partial c_{\mathrm{i}\alpha}}\right|_{\{c_\alpha\}} \Delta c_{\mathrm{i}\alpha} \\ &+ \frac{1}{2} \sum_{\substack{\mathrm{i}\alpha \\ \mathrm{j}\alpha'}} \left.\frac{\partial^2 F}{\partial c_{\mathrm{i}\alpha} \partial c_{j\alpha'}}\right|_{\{c_\alpha\}} \Delta c_{\mathrm{i}\alpha} \Delta c_{\mathrm{j}\alpha'} + \ldots \end{aligned} \tag{2.16}$$

to second-order.

---

[4] We allow the possibility of a functional, *i.e.*, a function of a function, for the case that the concentrations are a function of some other variable(s), *e.g.*, temperature.

[5] The host-dependent setting can be more convenient as any considered chemical fluctuation in this setting naturally preserves $\sum_\alpha c_{\mathrm{i}\alpha} = 1$.

Because all sites are crystallographically equivalent in the reference state (a condition which we imposed earlier), the first derivative of the free energy, $\frac{\partial F}{\partial c_{\mathrm{i}\alpha}}\big|_{\{c_\alpha\}}$, must be independent of its site label, i. In addition, to conserve overall concentrations, $\sum_{\mathrm{i}} \Delta c_{\mathrm{i}\alpha} = 0$ for all species $\alpha$. The first order term in 2.16 therefore vanishes. The lowest-order term in the series expansion which is non-zero is therefore the second-order term.

Proceeding, we use the Fourier-transformed site-wise concentrations to write the change in free energy due to an imposed fluctuation $\Delta c_{a\alpha}(\mathbf{k})$ as

$$\Delta F[\{c_{a\alpha}(\mathbf{k})\}] = \frac{1}{2} \sum_{a\alpha;b\alpha'} \sum_{\mathbf{k}} \Delta c_{a\alpha}(\mathbf{k}) F^{(2)}_{a\alpha;b\alpha'}(\mathbf{k}) \Delta c_{b\alpha'}(\mathbf{k}), \tag{2.17}$$

where $F^{(2)}_{a\alpha;b\beta}(\mathbf{k})$ is an appropriately defined matrix. This matrix is diagonal in $\mathbf{k}$ as a result of the translational symmetry of the lattice. While this matrix is positive-definite,[6] the system is stable to chemical fluctuations about the homogeneous reference state. It is intuitive to think of the homogeneous reference state as sitting in the (local) minimum of a bowl on the free energy surface of the alloy in this scenario.

A second-order phase transition occurs when, at some finite temperature, $T_{\mathrm{us}}$, the Hessian matrix $F^{(2)}_{a\alpha;b\alpha'}(\mathbf{k})$ stops being positive definite, *i.e.*, an eigenvalue is equal to or less than zero for some mode $\mathbf{k}_{\mathrm{us}}$ and some associated eigenvector.[7] This indicates that there is some chemical fluctuation $\Delta c_{a\alpha}(\mathbf{k})$ which, when applied to the disordered, homogeneous state, gives a state which is of the same energy or lower energy than the homogeneous state. The exact temperature at which the transition occurs is obtained by considering the temperature for which the lowest-lying eigenvalue of the matrix passes through zero. Formally, the (partial) ordering established, $\Delta c_{a\alpha}$, the temperature at which it occurs, $T_{\mathrm{us}}$, and the wavevector describing the ordering, $\mathbf{k}_{\mathrm{us}}$, are found by solving

$$\min_{\mathbf{k},\sigma} \left( \lambda_\sigma \left[ F^{(2)}_{a\alpha;b\alpha'}(\mathbf{k}) \right] \right) = 0, \tag{2.18}$$

where $\lambda_\sigma[M]$ denotes the $\sigma^{\mathrm{th}}$ eigenvalue of a matrix $M$. The $\Delta c_{a\alpha}$ is then the associated eigenvector.

That the disorder-order transition characterised by the wavevector $\mathbf{k}_{\mathrm{us}}$ is second-order is not guaranteed; to ascertain whether a transition is first- or second-order[8] we need information about higher-order derivatives of the free energy [14–16]. The third-order term in the series expansion is written

$$\frac{1}{3!} \sum_{a\alpha;b\alpha';c\alpha''} \sum_{\mathbf{k}_1,\mathbf{k}_2,\mathbf{k}_3} F^{(3)}_{a\alpha;b\alpha';c\alpha''}(\mathbf{k}_1, \mathbf{k}_2, \mathbf{k}_3) \Delta c_{a\alpha}(\mathbf{k}_1) \Delta c_{b\alpha'}(\mathbf{k}_2) \Delta c_{c\alpha''}(\mathbf{k}_3), \tag{2.19}$$

---

[6] A positive definite matrix is a matrix with all eigenvalues strictly positive.

[7] The importance of these eigenvectors will be discussed presently.

[8] Recall that a first-order transition is characterised by a discontinuity in order parameter through the phase transition, while a second-order transition has a continuous order parameter.

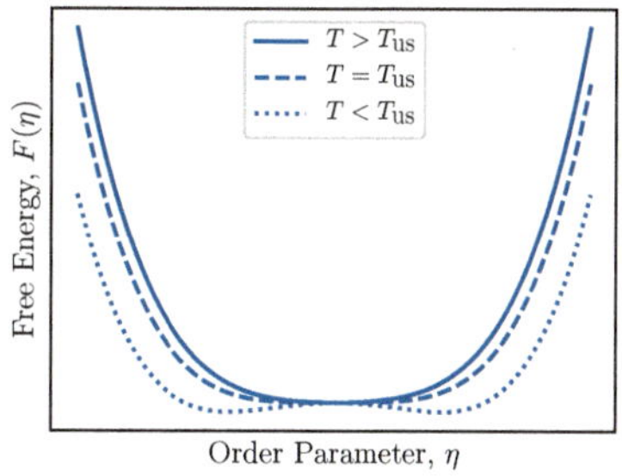


(a) Second-Order.

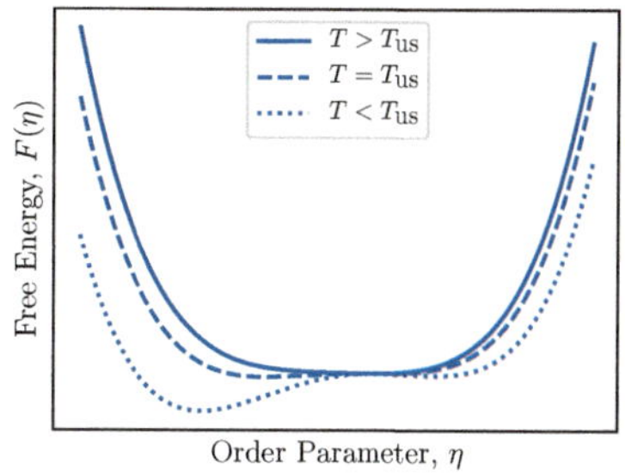


(b) First-Order.

**Fig. 2.3** Visualisations of toy, 1D free energies for the case of second- and first-order phase transitions. The second order case has the third derivative of the free energy set to zero, while the first order one has finite third derivative

while the fourth-order term is written similarly. Here, the $\{\mathbf{k}_i\}$ are vectors in the star of $\mathbf{k}_{\rm us}$, where the star of a vector means that vector and all other vectors which differ from it by a reciprocal lattice vector, *i.e.*, those which are crystallographically equivalent. If the third-order term is non-vanishing, or if the fourth-order term is negative, the disorder-order transition is first-order and may occur at some higher temperature. It has been shown by Khachaturyan [9, 10] that a necessary and sufficient condition for a transition to be second-order is that $\mathbf{k}_1 + \mathbf{k}_2 + \mathbf{k}_3 \neq \mathbf{G}$, where $\mathbf{G}$ is a reciprocal lattice vector, for any vectors $\{\mathbf{k}_i\}$ in the star of $\mathbf{k}_{\rm us}$. Even in the case of a first-order transition, the second-order term vanishing marks an absolute instability point—the system is unstable to any (arbitrarily small) fluctuation. Visual illustrations of a toy free energy in the cases of phase stability, second- and first-order phase transitions are given in Fig. 2.3. In the case of a first-order transition we will also see that hope is not lost, as the derivatives of the free energy can be mapped directly to an atomistic model from which both first- and second-order phase transitions can be inferred directly.

## 2.4 The Free Energy of an Alloy

The approach we are about to outline takes advantage of the grand canonical ensemble, for which the appropriate thermodynamic potential is the Landau free energy (also referred to as the grand potential), which is a function(al) of what we will refer to as site-wise chemical potentials, $\nu_{{\rm i}\alpha}$, describing the energetic cost of introducing an atom of species $\alpha$ at site i. This free energy, $\Omega$, can be obtained from the partition function via

$$\Omega = -\frac{1}{\beta} \ln Z \tag{2.20}$$

$$= -\frac{1}{\beta} \ln \sum_{\{\xi_{i\alpha}\}} e^{-\beta(E_{\text{int}}[\{\xi_{i\alpha}\}]+\sum_{i\alpha} \nu_{i\alpha}\xi_{i\alpha})}, \tag{2.21}$$

where $\beta = \frac{1}{k_B T}$, $Z$ denotes the partition function, and $E_{\text{int}}$ is the internal energy of the system, which is a function(al) of the site-wise occupancies.

Note that the free energy of 2.21 is invariant under a gauge transformation $\nu_{i\alpha} \mapsto \nu_{i\alpha} + \gamma_i$ because such a transformation leaves the probabilities unchanged. We fix this gauge degree of freedom by specifying that $\nu_{is} = 0$, where $s$ is the index of the host species. Note also that, in its current form, $\Omega = \Omega(T, \nu_{i\alpha})$, and does not explicitly depend on the site-wise concentrations. However, the $\nu_{i\alpha}$ are closely related to the $c_{i\alpha}$ because, by definition,

$$\langle \xi_{i\alpha} \rangle [\{\nu_{i\alpha}\}] = c_{i\alpha}, \tag{2.22}$$

as the on-site chemical potentials are chosen precisely to constrain the site-wise concentrations. It is therefore insignificant whether we minimise $\Omega(T, \nu_{i\alpha})$ subject to the constraint $\frac{1}{N}\sum_i c_{i\alpha} = c_\alpha$, or whether we minimise $\Omega(T, c_{i\alpha})$ with unknown $\nu_{i\alpha}$ subject to the same constraint.

### 2.4.1 Variational Free Energy

It remains to determine an explicit form of $\Omega(T, \{c_{i\alpha}\})$ which is computationally tractable. While $E_{\text{int}}[\{\xi_{i\alpha}\}]$ is, in principle, accessible to calculations using energies evaluated via density functional theory (DFT) or through appropriate interatomic potentials, the size of the space of possible configurations pushes evaluation of the free energy directly for any reasonably sized system beyond the realms of computational tractability, and approximations must be made.

We start by choosing to model the probability distribution for the system as a perturbation from an uncorrelated reference distribution, $P_0[\{\xi_{i\alpha}\}] = \prod_i \mathcal{P}_0(\xi_{i\alpha})$, *i.e.*, the probability distribution for a given site is independent of the distributions for surrounding sites. We write the site-wise concentrations $\bar{c}_{i\alpha} = \mathcal{P}_0(\xi_{i\alpha})$, with the bar indicating that these are generated by the uncorrelated distribution and are arbitrary at this point. (Such an uncorrelated distribution could be generated by a mean-field Hamiltonian, such as one of the form $H_{\text{MF}}(\{\nu_{i\alpha}\}) = \sum_{i\alpha} \bar{V}_{i\alpha}\xi_{i\alpha}$.) Let $H_0$ denote the Hamiltonian generating the uncorrelated distribution. Expanding 2.21 to first order in $\beta(H - H_0)$, we have that

$$\Omega = -\frac{1}{\beta} \ln \sum_{\{\xi_{i\alpha}\}} e^{-\beta(E_{\text{int}}-H_0)-\beta(H_0-\sum_{i\alpha} \nu_{i\alpha}\xi_{i\alpha})} \tag{2.23}$$

$$\approx -\frac{1}{\beta} \ln \sum_{\{\xi_{i\alpha}\}} (1 - \beta(E_{\text{int}} - H_0)) e^{-\beta(H_0 - \sum_{i\alpha} \nu_{i\alpha} \xi_{i\alpha})} \tag{2.24}$$

$$\approx \Omega_0 + \langle E_{\text{int}} - H_0 \rangle_0. \tag{2.25}$$

We note that this expansion is most valid for small values of $\beta$ and weakly correlated systems in which $(E_{\text{int}} - H_0)$ is 'small' in the appropriate sense.

By the Gibbs-Bogoliubov-Feynman inequality [17], this is an upper bound on the free energy of the system. Writing $\Omega_0 = \langle H_0 \rangle_0 - T\langle S_0 \rangle_0$, the inequality takes the form

$$\begin{aligned} \Omega &\leq \langle E_{\text{int}} - H_0 \rangle_0 + \Omega_0 \\ &= \langle E_{\text{int}} \rangle_0 - T\langle S_0 \rangle_0 \\ &=: \Omega^{(1)}. \end{aligned} \tag{2.26}$$

Because the entropy of the uncorrelated system is known (it is simply the so-called entropy of mixing [18]) our upper bound on the free energy of the system is then

$$\Omega^{(1)} = \langle H \rangle_0 + \beta^{-1} \sum_{i\alpha} \bar{c}_{i\alpha} \ln \bar{c}_{i\alpha} - \sum_{i\alpha} \nu_{i\alpha} \bar{c}_{i\alpha}, \tag{2.27}$$

where we re-iterate that the bar denotes that these are the site-wise concentrations generated by the uncorrelated probability distribution, which approximate the true site-wise concentrations of the system.

We can optimise the choice of parameters used in our mean field Hamiltonian by minimising $\Omega^{(1)}$ with respect to the site-wise concentrations $\{\bar{c}_{i\alpha}\}$ using the variational principle

$$\begin{aligned} 0 &= \frac{\partial \Omega^{(1)}}{\partial \bar{c}_{i\alpha}} \\ &= \frac{\partial \langle H \rangle_0}{\partial \bar{c}_{i\alpha}} + \beta^{-1} \ln \frac{\bar{c}_{i\alpha}}{\bar{c}_{is}} - \nu_{i\alpha}. \end{aligned} \tag{2.28}$$

In turn, this can be shown to be imply that

$$\bar{c}_{i\alpha} = \frac{\exp\left(-\beta \frac{\partial \langle E_{\text{int}} \rangle_0}{\partial \bar{c}_{i\alpha}} - \nu_{i\alpha}\right)}{1 + \sum_{\alpha}' \exp\left(-\beta \frac{\partial \langle E_{\text{int}} \rangle_0}{\partial \bar{c}_{i\alpha}} - \nu_{i\alpha}\right)}. \tag{2.29}$$

explicitly defining the site-wise concentrations when the average is taken with respect to the ensemble generated by the mean field Hamiltonian.

In summary, we now have an explicit form for our mean-field free energy (2.27) with optimal parameters determined variationally (2.28). It remains to perform the

Taylor expansion of the Landau theory and assess the response of the system to a perturbation of the on-site concentrations.

## 2.5 Energetic Cost of Fluctuations

We now assess the system's response to perturbations from the homogeneous reference state by expanding the mean-field free energy of 2.27 in terms of $\Delta\bar{c}_{i\alpha}$. This will, in fact, also lead us to a direct estimate of the atomic short-range order (ASRO), denoted $\Psi_{i\alpha;j\alpha'}$. We define this quantity as

$$\Psi_{i\alpha;j\alpha'} \equiv \langle\xi_{i\alpha}\xi_{j\alpha'}\rangle - \langle\xi_{i\alpha}\rangle\langle\xi_{j\alpha'}\rangle. \tag{2.30}$$

This is a measure of second-order correlations between lattice site occupancies, which is the fundamental quantity assessed by the diffuse scattering intensity in a diffraction experiment on an alloy. There are also now a number of more direct experimental techniques capable of measuring this quantity, too [19].

The atomic short-range order is directly accessible by considering derivatives of the alloy free energy. Note that from 2.21 we have that

$$-\frac{\partial\Omega}{\partial\nu_{i\alpha}} = \langle\xi_{i\alpha}\rangle = c_{i\alpha}. \tag{2.31}$$

Differentiating the same expression again gives

$$-\frac{\partial^2\Omega}{\partial\nu_{i\alpha}\partial\nu_{j\alpha'}} = \frac{\partial c_{i\alpha}}{\partial\nu_{j\alpha'}} \tag{2.32}$$

$$= \beta\Psi_{i\alpha;j\alpha'}, \tag{2.33}$$

where 2.33 also follows from 2.21 after a few lines of algebra. The relationship between the derivatives of the site-wise concentrations and the short-range order is particularly useful, because it means that we can estimate the ASRO from our optimal variational parameters via

$$\bar{\Psi}_{i\alpha;j\alpha'} = \beta^{-1}\frac{\partial\bar{c}_{i\alpha}}{\partial\nu_{j\alpha'}}, \tag{2.34}$$

where the bar serves as a reminder that this is an estimate of the true ASRO.

Now, differentiating 2.28 with respect to $\bar{c}_{j\alpha'}$ while holding other concentrations fixed, we find

$$0 = \frac{\partial^2\langle E_{\text{int}}\rangle_0}{\partial\bar{c}_{i\alpha}\partial\bar{c}_{j\alpha'}} + \beta^{-1}\left(\delta_{ij}\delta_{\alpha\alpha'}\frac{1}{\bar{c}_{j\alpha'}} + \delta_{ij}\frac{1}{\bar{c}_{is}}\right) - \frac{\partial\nu_{i\alpha}}{\partial\bar{c}_{j\alpha'}}, \tag{2.35}$$

or

$$0 = \beta^{-1}\left(\delta_{\mathrm{ij}}\delta_{\alpha\alpha'}\frac{1}{\bar{c}_{\mathrm{j}\alpha'}} + \delta_{\mathrm{ij}}\frac{1}{\bar{c}_{\mathrm{i}s}} - (\bar{\Psi}_{\mathrm{i}\alpha;\mathrm{j}\alpha'})^{-1}\right) - \frac{\partial^2\langle E_{\mathrm{int}}\rangle_0}{\partial\bar{c}_{\mathrm{i}\alpha}\partial\bar{c}_{\mathrm{j}\alpha'}}. \tag{2.36}$$

This means we can use the quantity $\frac{\partial^2\langle E_{\mathrm{int}}\rangle_0}{\partial\bar{c}_{i\alpha}\partial\bar{c}_{j\alpha'}}$ to directly estimate the ASRO. This is the central quantity of the theory, and for notational convenience we define

$$S^{(2)}_{\mathrm{i}\alpha;\mathrm{j}\alpha'} := \left.\frac{\partial^2\langle E_{\mathrm{int}}\rangle_0}{\partial\bar{c}_{\mathrm{i}\alpha}\partial\bar{c}_{\mathrm{j}\alpha'}}\right|_{\bar{c}_{\mathrm{i}s}\,\text{dependent}}, \tag{2.37}$$

$$\tilde{S}^{(2)}_{\mathrm{i}\alpha;\mathrm{j}\alpha'} := \left.\frac{\partial^2\langle E_{\mathrm{int}}\rangle_0}{\partial\bar{c}_{\mathrm{i}\alpha}\partial\bar{c}_{\mathrm{j}\alpha'}}\right|_{\bar{c}_{\mathrm{i}s}\,\text{independent}}. \tag{2.38}$$

As shorthand these quantities will often be referred to as 'the $S^{(2)}$s,' where the '2' in the superscript is to denote that these come from a second derivative. Note that 2.37 defines these quantities in the host-dependent setting, while 2.38 defines them in the host independent setting.

Next, we Taylor expand our expression for the mean field free energy, 2.27. The 0th order term is known, the 1st order term is zero, and the 2nd derivative of the mean-field free energy can be shown to be

$$\begin{aligned}\left.\frac{\partial^2\Omega^{(1)}}{\partial\bar{c}_{\mathrm{j}\alpha'}\partial\bar{c}_{\mathrm{k}\alpha''}}\right|_{\{c_\alpha\}} &= \frac{\partial\nu_{\mathrm{k}\alpha''}}{\partial\bar{c}_{\mathrm{j}\alpha'}} - \sum_{\mathrm{i}\alpha}{}' \frac{\partial^2\nu_{\mathrm{i}\alpha}}{\partial\bar{c}_{j\alpha'}\partial\bar{c}_{\mathrm{k}\alpha''}}\bar{c}_{\mathrm{i}\alpha} \\ &= \beta^{-1}\bar{\Psi}^{-1}_{\mathrm{j}\alpha';\mathrm{k}\alpha''} - \sum_{\mathrm{i}\alpha}{}' \frac{\partial^2\nu_{\mathrm{i}\alpha}}{\partial\bar{c}_{\mathrm{j}\alpha'}\partial\bar{c}_{\mathrm{k}\alpha''}}\bar{c}_{\mathrm{i}\alpha}.\end{aligned} \tag{2.39}$$

The second term relates to the change in chemical potentials as a result of the change in the $\{\bar{c}_{\mathrm{i}\alpha}\}$. If we can view $\{\nu_{\mathrm{i}\alpha}\}$ fixed and independent of $\{\bar{c}_{\mathrm{i}\alpha}\}$ this term would be zero. Viewing the $\{\nu_{\mathrm{i}\alpha}\}$ as a function of the concentrations is equivalent to working in the canonical ensemble. Our expectation is that in the thermodynamic limit, these variations are not important and do not impact the relevant physics. We therefore set them to zero and write $\Delta\Omega^{(1)} = \Delta F^{(1)}$.

Now, performing a second order expansion of the expression for the mean field free energy (2.27) we find

$$\Delta F^{(1)} = \frac{1}{2}\sum_{\substack{\mathrm{i}\alpha\\ \mathrm{j}\alpha'}}{}' \Delta\bar{c}_{\mathrm{i}\alpha}\left(\beta^{-1}\bar{\Psi}^{-1}_{\mathrm{i}\alpha;\mathrm{j}\alpha'}\right)\Delta\bar{c}_{\mathrm{j}\alpha'} + \ldots, \tag{2.40}$$

where the next term in the expansion is third-order or higher. Taking the Fourier transform and keeping only terms to second order, we have by our earlier result

$$\Delta F^{(1)} = \frac{1}{2} \sum_{\mathbf{k}} \sum_{\substack{a\alpha \\ b\alpha'}}{}' \Delta \bar{c}_{a\alpha}(\mathbf{k}) \left[\beta^{-1} \bar{\Psi}^{-1}_{a\alpha;b\alpha'}(\mathbf{k})\right] \Delta \bar{c}_{b\alpha'}(\mathbf{k}) \tag{2.41}$$

$$= \frac{1}{2} \sum_{\mathbf{k}} \sum_{\substack{a\alpha \\ b\alpha'}}{}' \Delta \bar{c}_{a\alpha}(\mathbf{k}) \left[(\beta C)^{-1}_{\alpha\alpha'} \delta_{ab} - S^{(2)}_{a\alpha;b\alpha'}(\mathbf{k})\right] \Delta \bar{c}_{b\alpha'}(\mathbf{k}), \tag{2.42}$$

where we define

$$C_{\alpha\alpha'} = c_\alpha (\delta_{\alpha\alpha'} - c_{\alpha'}), \tag{2.43}$$

and

$$C^{-1}_{\alpha\alpha'} = \left( \frac{\delta_{\alpha\alpha'}}{c_\alpha} + \frac{1}{c_s} \right). \tag{2.44}$$

Alternatively, in the host independent setting, these quantities are written

$$\Delta F^{(1)} = \frac{1}{2} \sum_{\mathbf{k}} \sum_{\substack{a\alpha \\ b\alpha'}} \Delta \bar{c}_{a\alpha}(\mathbf{k}) \left[(\beta \tilde{C})^{-1}_{\alpha\alpha'} \delta_{ab} - \tilde{S}^{(2)}_{a\alpha;b\alpha'}(\mathbf{k})\right] \Delta \bar{c}_{b\alpha'}(\mathbf{k}), \tag{2.45}$$

where

$$\tilde{C}^{-1}_{\alpha\alpha'} = \frac{\delta_{\alpha\alpha'}}{c_\alpha}. \tag{2.46}$$

The term in the square brackets we will refer to as the 'chemical stability matrix,' as it is the eigenvalues of this matrix that determine the stability of the solid solution to chemical fluctuations. Recalling that our variational free energy takes the form

$$\begin{aligned} F^{(1)} &= -TS + U \\ &= -\beta^{-1} \sum_{\mathrm{i}\alpha} \bar{c}_{\mathrm{i}\alpha} \ln \bar{c}_{\mathrm{i}\alpha} + \langle E_{\mathrm{int}} \rangle_0, \end{aligned} \tag{2.47}$$

we identify the first term in square brackets in 2.42 as the entropic cost of a variation. This is intuitive, as establishing atomic order lowers the entropy of the system. Then we identify the second term in square brackets as defining the energetic cost of establishing atomic order. In addition, from 2.41, it can be seen that the cost of a chemical fluctuation with mode $\mathbf{k}$ is, in a sense, inversely proportional to the atomic short-range order parameter. This is also intuitive, because a high value of the ASRO parameter implies that that correlation is likely and, therefore, its energetic cost cannot be particularly high.

We see, therefore, that by inverting the chemical stability matrix we can estimate the ASRO directly. In addition, as discussed in Sect. 2.3, the chemical stability matrix quantifies the stability of the solid solution to chemical fluctuations. When an eigenvalue of this matrix passes through zero at some temperature $T_{\mathrm{us}}$ for some mode $\mathbf{k}_{\mathrm{us}}$ and with associated eigenvector $\Delta c_\alpha$, we infer a phase transition.

## 2.6 Chemical Polarisations

Computationally, it is often convenient to work in the *host dependent* setting, where we consider the set of variables $(c_{\mathrm{i}1}(\mathbf{k}), \ldots, c_{\mathrm{i}(s-1)}(\mathbf{k}))$, *i.e.*, we have set species $s$ to be the host. This means that the chemical stability matrix is of dimension $(s-1) \times (s-1)$. To infer the stability of the solid solution, we required evaluation of the eigenvalues and eigenvectors of this matrix. However, this diagonalisation assumes the metric $\|\Delta\bar{c}\|^2 = \sum'_\alpha \Delta\bar{c}^2_{\mathrm{i}\alpha}$, which is dependent on the choice of host. The end result is that the obtained eigenvalues and eigenvectors have no intuitive physical meaning.

A better choice of metric is given by the sum over *all* components, $\|\Delta\bar{c}\|^2 = \sum_\alpha \Delta\bar{c}^2_{\mathrm{i}\alpha}$ (no prime on sum), which is not dependent on the choice of host and, therefore, gives a meaningful measure of the size of a chemical fluctuation. This requires working with the complete set of variables $(c_{\mathrm{i}1}(\mathbf{k}), \ldots, c_{\mathrm{i}s}(\mathbf{k}))$ in the *host independent* setting. In this setting, the chemical stability matrix is then of dimension $(s-1) \times (s-1)$. A problem arises, however, when this chemical stability matrix is diagonalised, because the additional degree of freedom allows eigenvectors that do not preserve $\sum_\alpha \Delta c_\alpha(\mathbf{k}) = 0$. To restore this constraint, the scheme proposed by Khan, Staunton and Stocks [7] is to first perform a change of variables which conserves the norm but isolates the non-physical degree of freedom. This transformation is given by

$$\Delta\eta_\mu(\mathbf{k}) = \sum_{\mu'} O_{\mu\mu'} \Delta c_{\mu'}(\mathbf{k}), \tag{2.48}$$

where $\Delta\eta_\mu$ represent the transformed variables, and the orthogonal transformation matrix, $O$, is defined as

$$O := \begin{pmatrix} 1/\sqrt{2}\,(1 & -1 & 0 & 0 & 0 & \cdots & 0) \\ 1/\sqrt{6}\,(1 & 1 & -2 & 0 & 0 & \cdots & 0) \\ 1/\sqrt{12}\,(1 & 1 & 1 & -3 & 0 & \cdots & 0) \\ \vdots & \vdots & \vdots & \vdots & \vdots & \ddots & \vdots \\ 1/\sqrt{s(s-1)}\,(1 & 1 & 1 & 1 & 1 & \cdots & 1-s) \\ 1/\sqrt{s}\,(1 & 1 & 1 & 1 & 1 & \cdots & 1) \end{pmatrix}. \tag{2.49}$$

The last row represents the sum $\Delta\eta_s = 1/\sqrt{s}\sum_\mu \Delta c_\mu$, which we require to be set to zero to ensure that considered fluctuations conserve the overall concentration of each species.

To construct the chemical stability matrix in this new set of variables, we transform the relevant quantities as

$$\tilde{S}^{(2)}_{\alpha\alpha'}(\mathbf{k}) \rightarrow \sum_{\mu\mu'} O_{\alpha\mu} \tilde{S}^{(2)}_{\mu\mu'}(\mathbf{k}) O^T_{\mu'\alpha'} =: \breve{S}^{(2)}_{\alpha\alpha'}(\mathbf{k})$$
$$(\beta\tilde{C})^{-1}_{\alpha\alpha'} \rightarrow \sum_{\mu\mu'} O_{\alpha\mu} (\beta\tilde{C})^{-1}_{\mu\mu'} O^T_{\mu'\alpha'} =: (\beta\breve{C})^{-1}_{\alpha\alpha'}.$$

The fluctuation to the free energy is then written

$$\Delta F^{(1)} = \frac{1}{2} \sum_k \sum_{\alpha\alpha'} \Delta\eta_\alpha(\mathbf{k}) \left[ (\beta\breve{C})^{-1}_{\alpha\alpha'} - \breve{S}^{(2)}_{\alpha\alpha'}(\mathbf{k}) \right] \Delta\eta_{\alpha'}(\mathbf{k}), \tag{2.50}$$

where the term in the square brackets is the new chemical stability matrix. Before diagonalising this quantity to obtain the eigenvalues and eigenvectors, it is necessary to restrict the diagonalisation to the correct subspace, the one in which the overall concentration of each species is conserved by the fluctuation. This is achieved by first setting

$$\breve{S}^{(2)}_{\alpha s}(\mathbf{k}) = \breve{S}^{(2)}_{s\alpha}(\mathbf{k}) = (\beta\breve{C})^{-1}_{s\alpha} = (\beta\breve{C})^{-1}_{\alpha s} = 0 \tag{2.51}$$

for all values of $\alpha$, then performing the diagonalisation. After diagonalising, there will be an artificial eigenvector $1/\sqrt{s}(1, \cdots, 1)$ with associated eigenvalue 0, which is discarded. The physically meaningful eigenvectors are then recovered by applying the backwards transformation

$$\Delta c_\mu(\mathbf{k}) = \sum_{\mu'} O^T_{\mu\mu'} \Delta\eta_{\mu'}(\mathbf{k}), \tag{2.52}$$

which allows for clear interpretation of the physical meaning of an eigenvalue.

## 2.7 The Onsager Correction

So far, we have outlined a conventional mean-field treatment of the problem. This, however, is known to overestimate transition temperatures, because on-site sum rules are violated [7]. In the case of the spatially homogeneous solid solution, the true atomic short-range order parameters are known to obey the on-site sum rule

$$\begin{aligned} \Psi_{\mathrm{i}\alpha;\mathrm{i}\alpha'} &= \langle \xi_{\mathrm{i}\alpha} \xi_{\mathrm{i}\alpha'} \rangle - \langle \xi_{\mathrm{i}\alpha} \rangle \langle \xi_{\mathrm{i}\alpha'} \rangle \\ &= \sum_{\{\xi_{\mathrm{i}\alpha}\}} P[\{\xi_{\mathrm{i}\alpha}\}] \xi_{\mathrm{i}\alpha} \xi_{\mathrm{i}\alpha'} - c_{\mathrm{i}\alpha} c_{\mathrm{i}\alpha'} \\ &= c_{\mathrm{i}\alpha} \delta_{\alpha\alpha'} - c_{\mathrm{i}\alpha} c_{\mathrm{i}\alpha'} = C_{\alpha\alpha'}. \end{aligned} \tag{2.53}$$

One approach to remedy this and restore the appropriate sum rules in a linear response theory is due to Onsager [20]. From 2.36, it is possible to form a self-consistent expression for $\bar{\Psi}_{\mathrm{i}\alpha;\mathrm{j}\alpha'}$, by writing

$$\beta\bar{\Psi}_{\mathrm{i}\alpha;\mathrm{j}\alpha'} = \left[(\beta C)^{-1} - S^{(2)}\right]^{-1}_{\mathrm{i}\alpha;\mathrm{j}\alpha'} \tag{2.54}$$

$$= \left[(\beta C) + (\beta C)S^{(2)}(\beta C) + \cdots\right]_{\mathrm{i}\alpha;\mathrm{j}\alpha'}$$

$$= \left[(\beta C) + (\beta C)S^{(2)}(\beta\bar{\Psi})\right]_{\mathrm{i}\alpha;\mathrm{j}\alpha'}. \tag{2.55}$$

Given that we estimated the short-range order by

$$\bar{\Psi}_{\mathrm{i}\alpha;\mathrm{j}\alpha'} = \beta^{-1}\frac{\partial\bar{c}_{\mathrm{i}\alpha}}{\partial\nu_{\mathrm{j}\alpha'}}, \tag{2.56}$$

we can write 2.55 in terms of explicit variations $\{\Delta c_{\mathrm{i}\alpha}\}$ and $\{\Delta\nu_{\mathrm{i}\alpha}\}$ as

$$\Delta\bar{c}_{\mathrm{i}\alpha} = \sum_{\alpha'}{}'(\beta C)_{\alpha\alpha'}\Delta\nu_{\mathrm{i}\alpha'} + \sum_{\gamma;\mathrm{k}\gamma'}{}'(\beta C)_{\alpha\gamma}S^{(2)}_{\mathrm{i}\gamma;\mathrm{k}\gamma'}\Delta\bar{c}_{\mathrm{k}\gamma'}. \tag{2.57}$$

The second term in this equation clearly violates the on-site sum rule of 2.53. We therefore seek a correction, or 'self-reaction field', $\Delta c^{(\nu)}_{\mathrm{j}\alpha'}$, which is the variation of the concentration on site j when considering the variation of the on-site chemical potentials $\{\Delta\nu_{\mathrm{i}1}, \ldots, \Delta\nu_{\mathrm{i}(s-1)}\}$. The restoration of the on-site sum rules is achieved by considering the ansatz

$$\Delta\hat{c}_{\mathrm{i}\alpha} - \sum_{\alpha'}{}'(\beta C)_{\alpha\alpha'}\Delta\nu_{\mathrm{i}\alpha'} = \sum_{\gamma;\mathrm{k}\gamma'}{}'(\beta C)_{\alpha\gamma}S^{(2)}_{\mathrm{i}\gamma;\mathrm{k}\gamma'}(\Delta\hat{c}_{\mathrm{k}\gamma'} - \Delta\hat{c}^{(\nu)}_{\mathrm{k}\gamma'})$$

$$= \sum_{\gamma;\mathrm{k}\gamma'}{}'(\beta C)_{\alpha\gamma}S^{(2)}_{\mathrm{i}\gamma;\mathrm{k}\gamma'}(\Delta\hat{c}_{\mathrm{k}\gamma'} - \sum_{\mu\mu'}{}'\frac{\partial\hat{c}_{\mathrm{k}\gamma'}}{\partial\nu_{\mathrm{i}\mu'}}(\beta C)^{-1}_{\mu\mu'}\Delta\hat{c}_{\mathrm{i}\mu}). \tag{2.58}$$

The new notation $\Delta\hat{c}_{\mathrm{i}\alpha}$ is made to reflect that these concentration variations are defined via the on-site chemical potentials. If only the on-site term $\Delta\nu_{\mathrm{i}\alpha}$ varies, the second term in 2.58 vanishes, so this definition is consistent with 2.53, and the sum rule is restored at all temperatures.

This ansatz can be interpreted as shifting the $S^{(2)}_{\mathrm{i}\gamma;\mathrm{k}\gamma'}$ by a temperature-dependent constant by rewriting 2.58 as

$$S^{(2)}_{\mathrm{i}\gamma;\mathrm{k}\gamma'} \to S^{(2)}_{\mathrm{i}\gamma;\mathrm{k}\gamma'} - \sum_{\mathrm{l}\mu;\mu'}{}' S^{(2)}_{\mathrm{i}\gamma;\mathrm{l}\mu}(\beta\hat{\Psi}_{\mathrm{l}\mu;\mathrm{i}\mu'})(\beta C)^{-1}_{\mu'\gamma'}\delta_{\mathrm{ik}} =: S^{(2)}_{\mathrm{i}\gamma;\mathrm{k}\gamma'} - \Lambda_{\gamma\gamma'}\delta_{\mathrm{ik}}. \tag{2.59}$$

This defines the corrected short-range order $\beta\hat{\Psi}_{\mathrm{l}\mu;\mathrm{i}\mu'} := \partial\hat{c}_{\mathrm{l}\mu}/\partial\nu_{\mathrm{i}\mu'}$. It is perhaps most convenient to understand this quantity in reciprocal space. Taking the lattice Fourier transform of 2.58 we find the site-independent Onsager cavity field,

$$\Lambda_{\mu\mu'} = \frac{1}{V_{\mathrm{BZ}}} \int \mathrm{d}\mathbf{k} \sum_{\gamma\gamma'}{}' S^{(2)}_{\mu\gamma}(\mathbf{k}) \hat{\Psi}_{\gamma\gamma'}(\mathbf{k}) C^{-1}_{\gamma'\mu'}, \tag{2.60}$$

which extends the temperature range over which the linear response theory is valid and should (in principle) reduce transition temperatures compared to their purely mean-field estimates.

To apply 2.60, it is necessary to fix $S^{(2)}_{\alpha\alpha'}(\mathbf{k})$ beforehand. For given $S^{(2)}_{\alpha\alpha'}(\mathbf{k})$ we can solve for $\hat{\Psi}_{\alpha\alpha'}(\mathbf{k})$ using 2.54. Then it is possible to compute $\Lambda_{\alpha\alpha'}$ using 2.60. These therefore define a set of equations which must be solved self-consistently, in practice in an iterative manner.

## 2.8 An Effective Pairwise Model: The Bragg-Williams Hamiltonian

So far, we have not specified the form of $E_{\mathrm{int}}[\{\xi_{i\alpha}\}]$, which could, in principle, come from a variety of sources, such as DFT calculations or interatomic potentials. Very often such techniques are prohibitively expensive to evaluate for large supercells and render conventional exploration of the phase space using Monte Carlo techniques impractical.[9] We will see now that, in principle, the theory outlined above can be used to provide a simple, pairwise, Ising-like Hamiltonian. This circumvents the problems associated with first- and second-order phase transitions outlined earlier, and also enables exploration of the phase space below any initial order-disorder transition temperature.

The primary limitation of this pairwise model is that, below any order-disorder transition temperature, the assumptions of the linear response theory are no longer valid. In general, we expect the pairwise interaction parameters defined presently to exhibit some temperature dependence, but we will not concern ourselves with this in this work. Any pairwise interactions extracted in the high-temperature limit will be assumed to be valid across the entire temperature range. If this seems like a crude approximation, bear in mind that it is actually a significant improvement on a number of other approaches. For example, [21] simulates the phase behaviour of a bcc high-entropy alloy using nearest-neighbour interactions only extracted from formation enthalpies of binary compounds from [22]. These binary compounds do not necessarily form on a bcc lattice, nor do we expect interactions to be limited only to nearest-neighbours, but still the approach proves successful. In addition, we emphasise that the description of the internal energy of our proposed atomistic model is purely a pairwise one, with no higher-order terms present. In principle, however, the methodology is extensible to evaluation of higher-order correlations and the corresponding many-body interactions, akin to a cluster expansion. We suggest that one approach which could be followed to obtain these would be analogous to that used

[9] Generally, sampling techniques require repeated evaluations of internal energies to explore the phase space of a system.

in [23, 24], which obtained higher-order interactions in a magnetic setting within a similar first-principles framework to that of the present work.

Having discussed some of its limitations, let us now outline how we obtain a pairwise description of the system's internal energy. The essence of the argument is that the expansion of 2.27 is unaffected if $E_{\text{int}}[\{\xi_{\text{i}\alpha}\}]$ is replaced by an effective Hamiltonian $H_{\text{eff}}$ which mimics the true internal energy of the system. The simplest physically reasonable Hamiltonian for such a system is simply a sum of pairwise terms, conventionally known as the Bragg-Williams Hamiltonian [25, 26], which is written

$$H\left(\{\xi_{\text{i}\alpha}\}\right) = \frac{1}{2}\sum_{\substack{\text{i},\alpha\\ \text{j},\alpha'}} \tilde{V}_{\text{i}\alpha;\text{j}\alpha'}\, \xi_{\text{i}\alpha}\xi_{\text{j}\alpha'} \tag{2.61}$$

in the host-independent setting, or

$$H\left(\{\xi_{\text{i}\alpha}\}\right) = \frac{1}{2}\sum_{\substack{\text{i},\alpha\\ \text{j},\alpha'}}{}' \, V_{\text{i}\alpha;\text{j}\alpha'}\, \xi_{\text{i}\alpha}\xi_{\text{j}\alpha'} \tag{2.62}$$

in the host dependent setting.[10] (The factor of 1/2 is to eliminate double-counting.) This defines how to evaluate $E_{\text{int}}[\{\xi_{\text{i}\alpha}\}]$ for this model. The parameters $\tilde{V}_{\text{i}\alpha;\text{j}\alpha'}$ and $V_{\text{i}\alpha;\text{j}\alpha'}$ denote somehow a 'potential', 'interaction', or interchange energy between species $\alpha$ on site $i$ and species $\alpha'$ on site $j$. We can think of these as an effective pair interaction, describing the energy associated with an atom of species $\alpha$ on lattice site i interacting with an atom of species $\alpha'$ on lattice site j.

To evaluate $\langle E_{\text{int}}\rangle_0[\{c_{\text{i}\alpha}\}]$, we simply replace $\xi_{\text{i}\alpha}$ in 2.61 or 2.62 with $c_{\text{i}\alpha}$. For example, in the host-independent setting, we write

$$\langle E_{\text{int}}\rangle_0\left[\{c_{\text{i}\alpha}\}\right] = \langle H\rangle_0\left[\{c_{\text{i}\alpha}\}\right] = \frac{1}{2}\sum_{\substack{\text{i},\alpha\\ \text{j},\alpha'}} \tilde{V}_{\text{i}\alpha;\text{j}\alpha'}\, c_{\text{i}\alpha}c_{\text{j}\alpha'}, \tag{2.63}$$

*i.e.*, a sum over pair interactions, as before.

To be a valid substitution in 2.27, we require that its variation is equal to the variation of the true internal energy, *i.e.*, $\Delta\langle H\rangle_0 = \Delta\langle\tilde{H}\rangle_0 = \Delta\langle E_{\text{int}}\rangle_0$ for a considered fluctuation around the homogeneous reference state. The variation of the Bragg-Williams Hamiltonian is written

[10] Presently we will see how to relate the parameters between the two settings.

$$\Delta\langle H\rangle_0 = \frac{1}{2}\sum_{i\alpha;j\alpha'}{}' V_{i\alpha;j\alpha'}(\Delta\bar{c}_{i\alpha}\bar{c}_\beta + \bar{c}_\alpha\Delta\bar{c}_{j\alpha'} + \Delta\bar{c}_{i\alpha}\Delta\bar{c}_{j\alpha'}) \tag{2.64}$$

$$= \frac{1}{2}\sum_{i\alpha;j\alpha'}{}' V_{i\alpha;j\alpha'}\Delta\bar{c}_{i\alpha}\Delta\bar{c}_{j\alpha'} \tag{2.65}$$

$$= \frac{1}{2}\sum_{i\alpha;j\alpha'} \tilde{V}_{i\alpha;j\alpha'}\Delta\bar{c}_{i\alpha}\Delta\bar{c}_{j\alpha'} \tag{2.66}$$

for an expansion about the homogeneous reference state. The first-order terms vanish because of the crystallographic equivalence of all lattice sites and because $\sum_i \Delta\bar{c}_{i\alpha} = 0$ to conserve overall concentrations. 2.65 and 2.66 can be shown to relate the pairwise interactions between the host-dependent and host-independent settings by

$$V_{i\alpha;j\alpha'} = \tilde{V}_{i\alpha;j\alpha'} + \tilde{V}_{is;js} - \tilde{V}_{is;j\alpha'} - \tilde{V}_{i\alpha;js}. \tag{2.67}$$

The inverse map, from the host-dependent setting to the host-independent setting, $V_{i\alpha;j\alpha'} \to \tilde{V}_{i\alpha;j\alpha'}$, is not uniquely defined. This is because the expansion of 2.64 is unaffected by a gauge transformation $\tilde{V}_{i\alpha;j\alpha'} \to \tilde{V}_{i\alpha;j\alpha'} + \phi_\alpha\phi_\beta$ for any mean-field, *i.e.*, constant, term $\phi_\alpha$.

Fixing this gauge momentarily, by comparison with 2.27, we identify

$$V_{i\alpha;j\alpha'} = -\, S^{(2)}_{i\alpha;j\alpha'} \tag{2.68}$$

$$\tilde{V}_{i\alpha;j\alpha'} = -\, \tilde{S}^{(2)}_{i\alpha;j\alpha'}. \tag{2.69}$$

This means that the $S^{(2)}$s represent an *unambiguous* best choice of parameters to be used in the pairwise interactions of 2.61 or 2.62 under the assumptions stated.

To close this section, we will now define a map which inverts 2.67. For form of $\langle E_{\mathrm{int}}\rangle_0$ used in Chap. 3, $\tilde{S}^{(2)}_{i\alpha;j\alpha'}$ obeys the sum rule [7]

$$\sum_\alpha \bar{c}_{i\alpha}\tilde{S}^{(2)}_{i\alpha;j\alpha'} = 0 \tag{2.70}$$

for all $i\alpha; j\alpha'$. If this sum rule is obeyed, it means we can fix the gauge on $\tilde{V}_{i\alpha;j\alpha'}$ and define the reverse map from the host-dependent to host-independent setting, $V_{i\alpha;j\alpha'} \to \tilde{V}_{i\alpha;j\alpha'}$. This is defined by

$$\tilde{V}_{is;js} = \sum_{\alpha\alpha'}{}' \bar{c}_{i\alpha}\bar{c}_{j\alpha}V_{i\alpha;j\alpha'}$$

$$\tilde{V}_{is;j\alpha'} = \tilde{V}_{is;js} - \sum_{\alpha}{}' \bar{c}_{i\alpha}V_{i\alpha;j\alpha'}$$

$$\tilde{V}_{i\alpha;j\alpha'} = V_{i\alpha;j\alpha'} - \tilde{V}_{is;js} + \tilde{V}_{is;j\alpha'} + \tilde{V}_{is;j\alpha}.$$

It is generally more intuitive to work in the host-independent setting when considering the Bragg-Williams model for a multicomponent alloy, as one can intuitively identify the energetics associated with interactions between different chemical species. However, for computations of $S^{(2)}_{\alpha\alpha'}(\mathbf{k})$ as outlined in Chap. 3, it can be helpful to use the host-dependent setting. The maps defined above allow one to switch freely between the two settings.

## 2.9 Atomistic Modelling Approaches

The first advantage of the form of the internal energy of 2.61 or 2.62 is that, as a lattice-based, pairwise model, it is easy to interpret; we can inspect the fitted $\tilde{V}_{i\alpha j\alpha'}$ and intuitively see which pairs of chemical species are favoured/disfavoured, and the relative strength of interactions. The second advantage is that it lends itself to computationally cheap, atomistic simulations, of which two will be outlined in this chapter. The first works directly with the Bragg-Williams Hamiltonian using discrete site-occupancies, while the second works with the continuous, site-wise concentrations and will be referred to as the 'lattice gas' model [8].

The advantage of either approach is that, once $\tilde{S}^{(2)}_{\mathrm{i}\alpha;\mathrm{j}\alpha'}$ has been obtained, we have no further need to evaluate $E_{\mathrm{int}}[\{\xi_{\mathrm{i}\alpha}\}]$ or its derivatives. This is important because, very often, the form of $E_{\mathrm{int}}[\{\xi_{\mathrm{i}\alpha}\}]$ will be prohibitively expensive to evaluate repeatedly. The 'gold standard' for such energies is taken to be a DFT calculation which, for a reasonably large supercell, swiftly becomes highly computationally expensive. Indeed, even a machine-learned interatomic potential is significantly more computationally expensive to evaluate than the simple interaction of 2.62 or 2.61.

### *2.9.1 Representation and Fitting of Pairwise Interactions*

The notation used for the pairwise interactions, $\tilde{V}_{\mathrm{i}\alpha\mathrm{j}\alpha'}$, is not immediately intuitive. However, because the approach used to calculate $\tilde{S}^{(2)}_{\mathrm{i}\alpha;\mathrm{j}\alpha'}$ involved an expansion about a reference state that was both homogeneous and isotropic, $\tilde{V}_{\mathrm{i}\alpha\mathrm{j}\alpha'}$ can also be assumed to be homogeneous and isotropic. We therefore write $V^{(n)}_{\alpha\alpha'}$ to denote the interaction between species $\alpha$ and $\alpha'$ on coordination shell[11] $n$.

In addition, in what will follow, the linear response calculation will be performed in reciprocal space, and $S^{(2)}_{\alpha\alpha'}(\mathbf{k})$ will be sampled at a finite number of points of the irreducible Brillouin zone (IBZ). To recover the real-space interaction, it is necessary to backwards Fourier transform and fit the data to a function of the form

[11] By 'coordination shell $n$', we mean those lattice sites which are $n$th neighbours of a given lattice site.

$$S^{(2)}_{\alpha\alpha'}(\mathbf{k}) \approx -\sum_{n=0}^{N} V^{(n)}_{\alpha\alpha'} \left( \sum_{\{\mathbf{R}_i\}_n} e^{i\mathbf{k}\cdot\mathbf{R}_i} \right), \tag{2.71}$$

where $\{\mathbf{R}_i\}_n$ denotes the set of vectors pointing to all lattice sites on the $n$th neighbour shell, and $N$ denotes the maximum number of shells considered. The $V^{(n)}_{\alpha\alpha'}$ are the coefficients fitted.

The cut-off distance, corresponding to the distance to the furthest considered coordination shell $N$, need not be particularly large for the systems considered in this work. This is because we will always consider metallic systems for which the effects of electronic screening will be significant. Generally, on close-packed lattices (fcc, bcc, hcp) we find that between two and six coordination shells are needed to fit our reciprocal-space data to a real-space interaction with good accuracy.[12] This is naturally in accordance with the 'nearsightedness' principle proposed by Kohn and used as the basis of linear-scaling DFT calculations [27]. Broadly, this principle states that, in 'nice' physical systems, we expect the electron density at a given point to only be influenced by its local environment, and not by what is happening a 'long' distance away. In the context of the Bragg-Williams Hamiltonian, this naturally means that, for well-behaved systems, only interactions on the first few coordination shells are expected to make substantial contributions to the internal energy of the system.

### *2.9.2 Sampling Techniques for the Bragg-Williams Model*

Consider first the case where we work with discrete site-occupancies and the conventional Bragg-Williams Hamiltonian. In this work, we will elect to use the Metropolis Monte Carlo algorithm with Kawasaki dynamics to conserve overall concentrations of each chemical species [28]. This method has been used in the past with success to study the physics of alloy formation [21, 29].

The algorithm proceeds as follows: the occupation numbers are randomly initialised with the overall number of atoms of each species, defined by the overall concentrations, as the only restriction. A pair of atomic sites is selected at random, and the change in internal energy $\Delta H$ from swapping the site occupancies is calculated. (This pair of sites need not be nearest neighbours; lifting this restriction enables equilibrium to be achieved more quickly.) If the change in energy is negative ($\Delta H < 0$) the swap is accepted unconditionally, while if the change is positive ($\Delta H > 0$) the swap is accepted with acceptance probability $e^{-\beta\Delta H}$. To assess the configurational contribution to the specific heat capacity (SHC) of the system, we use the fluctuation-dissipation theorem [30]. At thermodynamic equilibrium, this theorem allows us to estimate the specific heat capacity as

[12] As examples, we note that the first two coordination shells for the fcc (bcc) lattice represent the closest 18 (14) lattice sites to the one being considered, while the first six coordination shells for the fcc (bcc) lattice represents the 86 (64) closest lattice sites.

$$C = \frac{1}{k_B T^2} \left( \langle E^2 \rangle - \langle E \rangle^2 \right), \tag{2.72}$$

which in practice means evaluating $\langle E^2 \rangle$ and $\langle E \rangle$ for a (reasonably large) number of decorrelated samples.

To quantify atomic ordering in our simulations, we focus primarily on atomic short-range order, as this can also be used to classify long-range ordered structures when enough coordination shells are considered. We generate the Warren-Cowley ASRO parameters [31, 32], $\alpha_n^{pq}$, adapted to the multicomponent alloy setting,

$$\alpha_n^{pq} = 1 - \frac{P_n^{pq}}{c_q}, \tag{2.73}$$

where $n$ refers to the $n$th coordination shell, $P_n^{pq}$ is the conditional probability of an atom of type $q$ neighbouring an atom of type $p$ on coordination shell $n$, and $c_q$ is the overall concentration of atom type $q$. When $\alpha_n^{pq} > 0$, $p$-$q$ pairs are disfavoured on shell $n$, while when $\alpha_n^{pq} < 0$ they are favoured. The value 0 corresponds to the ideal, random, solid solution.

Combining plots of the specific-heat capacity and the atomic short-range order parameters as a function of temperature for a given simulation enables the phase behaviour of a system to be understood in detail, even below any initial order/disorder transition temperature. Indeed, many of the systems in this work will exhibit multiple chemical orderings occurring at separate temperatures.

### 2.9.3 *The Lattice Gas Model*

An alternative to working with the discrete, site occupancies and considering only atomic swaps, is to work instead with the site-wise concentrations, as in 2.63. The internal energy is written

$$H\left(\{c_{\mathrm{i}\alpha}\}\right) = \frac{1}{2} \sum_{\substack{\mathrm{i},\alpha \\ \mathrm{j},\alpha'}} \tilde{V}_{\mathrm{i}\alpha;\mathrm{j}\alpha'}\, c_{\mathrm{i}\alpha} c_{\mathrm{j}\alpha'} \tag{2.74}$$

in the host-independent setting, or

$$H\left(\{c_{\mathrm{i}\alpha}\}\right) = \frac{1}{2} \sum_{\substack{\mathrm{i},\alpha \\ \mathrm{j},\alpha'}}{}' \, V_{\mathrm{i}\alpha;\mathrm{j}\alpha'}\, c_{\mathrm{i}\alpha} c_{\mathrm{j}\alpha'} \tag{2.75}$$

in the host-dependent setting.

Rather than considering a sampling algorithm applied to this representation, in this work we instead use an approach due to Khachaturyan based on a gradient-descent algorithm [9, 10]. Although Khachaturyan originally proposed this approach to study the kinetics of alloy formation, it is also possible to use it to minimise the free energy of a model system and find its equilibrium configuration at a given temperature. This approach is particularly useful when systems form an intermetallic phase but are away from a stoichiometric composition. In such systems, this approach enables us to predict partial site-occupancies that can then be used in the KKR-CPA formulation of DFT to predict materials properties.

The free energy of the lattice gas is written as

$$F[\{c_{\mathrm{i}\alpha}\}] = -\beta^{-1} \sum_{\mathrm{i}\alpha} c_{\mathrm{i}\alpha} \ln c_{\mathrm{i}\alpha} + \frac{1}{2} \sum_{\substack{\mathrm{i},\alpha \\ \mathrm{j},\alpha'}} \tilde{V}_{\mathrm{i}\alpha;\mathrm{j}\alpha'}\, c_{\mathrm{i}\alpha} c_{\mathrm{j}\alpha'}. \tag{2.76}$$

At a given temperature below the ordering temperature of the system, the site-wise concentrations are initialised in some initial configuration (ideally a homogeneous distribution with a small amount of random noise to 'kick-start' the descent) and are evolved according to

$$\frac{\mathrm{d}c_{\mathrm{i}}}{\mathrm{d}t} = \sum_{\mathrm{j}} L_{\mathrm{ij}} \frac{\partial F}{\partial c_{\mathrm{j}}}, \tag{2.77}$$

where

$$L_{\mathrm{ij}} = \begin{cases} -N_{\mathrm{nn}}, & \mathrm{i} = \mathrm{j} \\ 1, & \mathrm{i}, \mathrm{j} \text{ nearest neighbours} \\ 0, & \text{otherwise} \end{cases} \tag{2.78}$$

to preserve overall concentration. $N_{\mathrm{nn}}$ denotes the number of nearest-neighbours on the lattice, *e.g.*, 8 in the case of the bcc lattice. We interpret a steady state solution, $\frac{dc_{\mathrm{i}}}{dt} = 0$ as the ground state configuration of the system for which the free energy is minimised. We will see an example of the application of this approach in Chap. 8.

## 2.10 Some Intuitive Examples

In what came before, we did not specify the form of the internal energy of the system used in the linear response analysis or to fit the pairwise interactions for atomistic modelling. This is to emphasise that the Landau theory is applicable, in principle, to any form of internal energy one can think of. The true internal energy of the alloy originates from a hugely complex system of interacting electrons and nuclei, and so it is necessary to make approximations on some level. We again emphasise that we could consider energies evaluated from first principles using DFT calculations, or we could consider some form of interatomic potential or cluster expansion.

To give insight, we will now study some intuitive examples where the internal energy is given exactly by the Bragg-Williams Hamiltonian for some choice of parameters $V_{i\alpha j\alpha'}$. We will illustrate results for both the linear response calculation and atomistic simulations for three example systems.

### 2.10.1 *Interactions and Chemical Stability Matrices Visualised*

First, let us consider how various interactions in this model might describe chemical orderings. Throughout these examples, we will use dimensionless units. First, we consider an example of an equiatomic binary system on the bcc lattice. To describe phase segregation in this system, it is sufficient to consider nearest-neighbour interactions, where like pairs are favoured, and unlike pairs are disfavoured. An interaction of the form

$$V^{(1)}_{\alpha\alpha'} = \begin{pmatrix} V/4 & -V/4 \\ -V/4 & V/4 \end{pmatrix} \tag{2.79}$$

gives $V^{(1)} = V$ when mapped back to the host-dependent setting. Constructing the chemical stability matrix and visualising its eigenvalues around the IBZ[13] of the bcc lattice in Fig. 2.4a, we see that the minimum eigenvalue lies at $\mathbf{k} = (0, 0, 0)$, *i.e.*, the $\Gamma$ point, indicative of phase segregation, as expected. The ordering temperature predicted within this mean field treatment is $k_B T_{\text{us}} = 2V$, in agreement with the well-known result for the Ising model. (A Bragg-Williams model phase-segregating is entirely mathematically equivalent to an Ising model for a ferromagnet [8].)

Switching the sign on the interaction will then describe a B2 ordering, where unlike nearest-neighbours are favoured. The eigenvalues of the chemical stability matrix around the IBZ are visualised in Fig. 2.4b. We see that the minimum eigenvalue is now at $\mathbf{k} = (0, 0, 1)$, *i.e.*, the $H$ point. Again, the critical temperature is $k_B T_{\text{us}} = 2V$.

Another illustrative example is given by considering a three-component alloy on the bcc lattice, with $c_\alpha = (\frac{1}{4}, \frac{1}{4}, \frac{1}{2})$. Our aim here will be to describe a Heusler-like ordering, *i.e.*, the $L2_1$ structure, visualised in Fig. 2.6. On the first shell, let us consider an interaction of the form

$$V^{(1)}_{\alpha\alpha'} = \begin{pmatrix} 0 & 0 & -V/4 \\ 0 & 0 & -V/4 \\ -V/4 & -V/4 & V/4 \end{pmatrix}, \tag{2.80}$$

which describes a case where species $A$ and $B$ do not interact on the first shell, while for species $C$ it is energetically favourable to have species $A$ and $B$ as nearest neighbours, and unfavourable to have itself as a neighbour. The eigenvalues for this case are visualised around the IBZ in Fig. 2.5a. A concentration wave mode similar to that of the binary B2 ordering can be seen, as can a new, flat mode, associated with the non-interacting species $A$ and $B$. At the $H$ point, the lowest lying eigenvalue

[13] A primer on the special points of the IBZ of the bcc and fcc lattices is given in Appendix A.

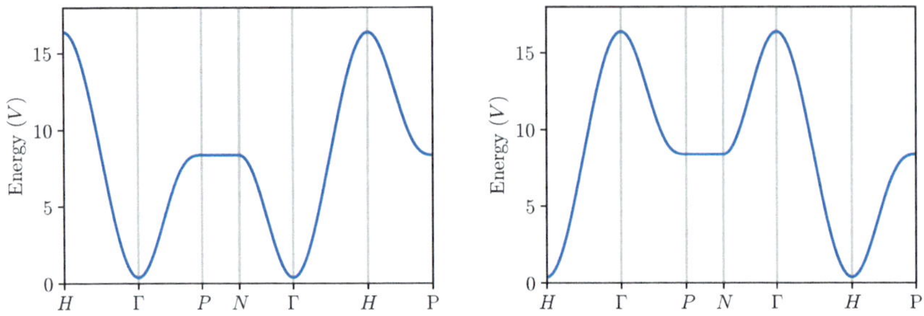


(a) Phase Segregation. (b) B2 Ordering.

**Fig. 2.4** Eigenvalues of the chemical stability matrix for a system on the bcc lattice with $c_\alpha = (1/2, 1/2)$ and interactions as specified in 2.79 (left) and the same interaction with the signs flipped (right) evaluated at $k_BT = 2.1V$. Phase segregation is indicated by the minimum at $\Gamma$, while a B2 ordering is indicated by the minimum at $H$

comes with associated eigenvector[14]

$$\Delta c_\alpha = (0.408,\ 0.408,\ -0.816), \tag{2.81}$$

while the eigenvalue at higher energy (associated with the flat mode) comes with eigenvector

$$\Delta c_\alpha = (0.707,\ 0.707,\ 0). \tag{2.82}$$

We see that the first mode is associated with an $(A,B;C)$ B2 ordering, while the second, flat mode, describes the non-interacting species $A$ and $B$. The mean-field ordering temperature in this case is now $k_BT_{\text{us}} = 1.5V$, reduced from the binary case, which is associated with an increased configurational entropy.

To lift the degeneracy and break the flat mode, consider now introducing an interaction on the second shell,

$$V^{(1)}_{\alpha\alpha'} = \frac{1}{5}\begin{pmatrix} V/4 & -V/4 & 0 \\ -V/4 & V/4 & 0 \\ 0 & 0 & 0 \end{pmatrix}, \tag{2.83}$$

which means that, on the second coordination shell, $A$-$B$ pairs are favoured, while $A$-$A$ and $B$-$B$ pairs are disfavoured. Note that this is scaled down in energy compared to the first shell, allowing the nearest-neighbour interaction to dominate the ordering. Plotting these eigenvalues in Fig. 2.5b, we see that the previously flat mode has acquired weak **k**-dependence, reflective of the strengthening second-shell interactions. The chemical polarisations for each mode remain as they were before, because the interactions on the two shells are, in a sense, linearly independent.

[14] Note that, in general, the chemical polarisations will change as we move around the IBZ. Here we specify their values at the $H$ point.

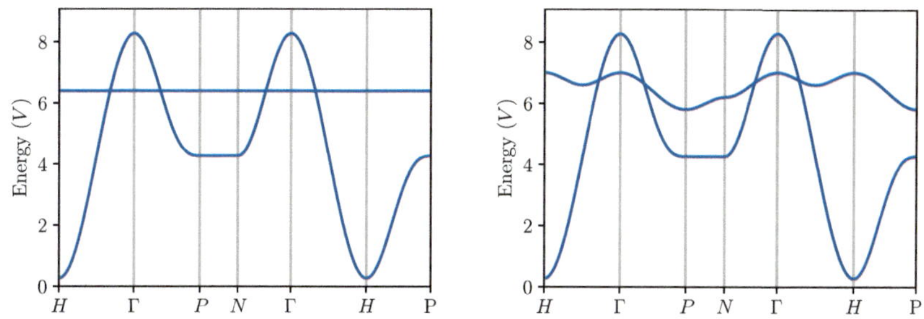


(a) Interactions as Eq. 2.80.

(b) Interactions as Eq. 2.83.

**Fig. 2.5** Eigenvalues of the chemical stability matrix for a Heusler system on the bcc lattice with $c_\alpha = (\frac{1}{4}, \frac{1}{4}, \frac{1}{2})$ and interactions as specified in 2.80 (left) and 2.83 (right) evaluated at $k_B T = 1.6V$. The B2 ordering is indicated by the minimum at $H$, while the weak interaction on the second shell introduced in **b** can be seen to lift the degeneracy of the flat mode

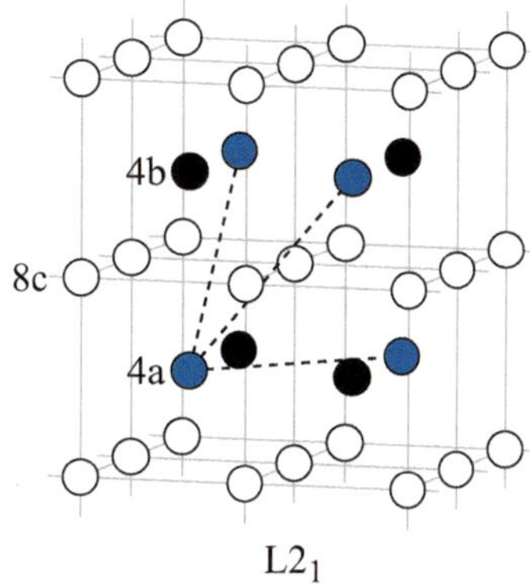


**Fig. 2.6** Visualisation of the $L2_1$ ordered structure imposed on the bcc lattice, which is the structure associated with Heusler compounds. These compounds have $c_\alpha = (\frac{1}{4}, \frac{1}{4}, \frac{1}{2})$

Similar examples can be constructed on the fcc lattice. Taking an interaction of the form

$$V^{(1)}_{\alpha\alpha'} = \begin{pmatrix} V/4 & -V/4 \\ -V/4 & V/4 \end{pmatrix} \tag{2.84}$$

gives the eigenvalue plot of Fig. 2.7a. Here it can be seen that the $X$ (corresponding to $L1_0$ or $L1_2$ order) and $W$ corresponding to $A_2B_2$ order are degenerate. The transition temperature is $k_B T_{\text{us}} = V$.

Adding an interaction on the second shell of the form

$$V^{(2)}_{\alpha\alpha'} = -\frac{1}{2}\begin{pmatrix} V/4 & -V/4 \\ -V/4 & V/4 \end{pmatrix} \tag{2.85}$$

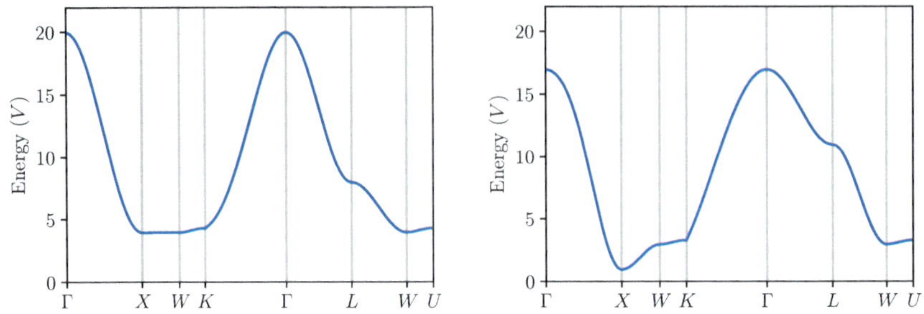


(a) Degeneracy between **X** and **W**. (b) Minimum at the **X** point.

**Fig. 2.7** Eigenvalues of the chemical stability matrix for a system on the fcc lattice with $c_\alpha = (0.5, 0.5)$ and interactions as specified in 2.84 (left) and 2.85 (right) evaluated at $k_B T = 1.6V$

lifts the degeneracy, as now on the second coordination shell *like* pairs are favoured, which corresponds to the $L1_0$ and $L1_2$ structures. The plots for the eigenvalues in this case can be seen in Fig. 2.7b. The ordering temperature with this additional interaction on the second coordination shell is now $k_B T_{\text{us}} = 1.75V$.

## 2.10.2 *Monte Carlo Simulations*

For the interactions outlined above, we also perform Monte Carlo simulations to further examine the phase behaviour of each system. Beginning with the B2 ordering, for which the eigenvalues of the chemical stability matrix were visualised in Fig. 2.4b, we perform an ensemble of 32 simulations on a cubic system of $8 \times 8 \times 8$ cubic unit cells for a total of 1024 atoms. The systems were prepared in an initial random configuration, then computationally annealed from high to low temperature. At each temperature, $10^4$ Monte Carlo steps per atom were performed to ensure equilibrium was achieved. To test convergence with respect to system size (*i.e.*, to address finite size effects) we perform simulations on a number of smaller system sizes and observe how quantities of interest evolve, while to assess the number of steps required to achieve equilibrium it is necessary to study the evolution of key quantities as a function of number of steps made.

The Warren-Cowley ASRO parameters as a function of temperature for the 1024 atom system are visualised in the left pane of Fig. 2.8. The B2 ordering, characterised by unlike pairs being favoured at nearest-neighbour distance and disfavoured at second-nearest-neighbour distance, can be seen to occur a little way below $k_B T = 2V$. This is to be expected; the mean-field analysis of the linear response theory is expected to overestimate transition temperatures. Then, in the right pane, we observe the convergence of the SHC curve with respect to system size. Explicitly, we consider systems of $8 \times 8 \times 8$, $6 \times 6 \times 6$, $4 \times 4 \times 4$, $3 \times 3 \times 3$, and $2 \times 2 \times 2$

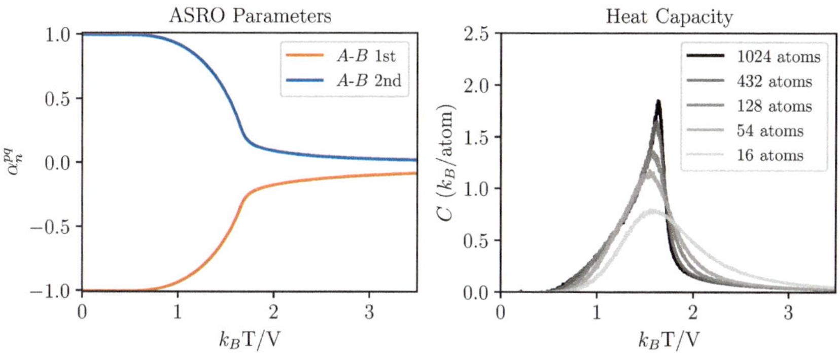


**Fig. 2.8** Visualisation of the Warren-Cowley ASRO parameters (left) and specific heat capacity (right) as a function of temperature for a Monte Carlo simulation of a system on the bcc lattice exhibiting a B2 ordering. The ASRO parameters are extracted from a simulation on $8 \times 8 \times 8$ cubic bcc unit cells, containing a total of 1024 atoms. The SHC is plotted for a variety of system sizes to demonstrate the convergence with respect to system size

cubic unit cells, for a total number of atoms of 1024, 432, 128, 64, and 16 atoms, respectively. It can be seen that the peak in SHC becomes narrower and shifts to lower temperatures with increasing system size, although by 1024 atoms the curve is clearly well-converged.

Proceeding to the three-component Heusler system, for which the eigenvalues of the chemical stability matrix were visualised in Fig. 2.5b, using the same parameters we perform Monte Carlo simulations and plot the Warren-Cowley ASRO parameters and SHC as a function of temperature in Fig. 2.9. Again, we perform an ensemble of 32 simulations on a cubic system of $8 \times 8 \times 8$ cubic unit cells for a total of 1024 atoms to extract the ASRO parameters, while the heat capacity is extracted from ensembles of simulations on a variety of system sizes to inspect its convergence with respect to system size. Note also that we do not consider the case of a $3 \times 3 \times 3$ cubic unit cell system, as this supercell is incommensurate with the $L2_1$ ordering.

When the Warren-Cowley ASRO parameters are considered in conjunction with the SHC plots, we see that the B2 ordering can be seen to occur at a marginally lower temperature than the mean-field estimate. In addition, however, we see that the interaction on the second coordination shell has introduced a second phase transition at lower temperatures. This transition is the transition from a B2 structure to full $L2_1$ order. This result emphasises that Monte Carlo simulations provide additional information to the linear response theory alone, by predicting potential possible further phase transitions below any initial ordering.

Finally, we consider the example of $L1_0$ ordering on the fcc lattice, for which the eigenvalues of the chemical stability matrix were visualised in Fig. 2.7b. Again, we consider a variety of system sizes, $8 \times 8 \times 8$, $6 \times 6 \times 6$, $4 \times 4 \times 4$, $3 \times 3 \times 3$, and $2 \times 2 \times 2$ cubic unit cells, which this time correspond to 2048, 864, 256, 108, and 32 atoms, owing to the fact that the cubic fcc unit cell contains four atoms. All other

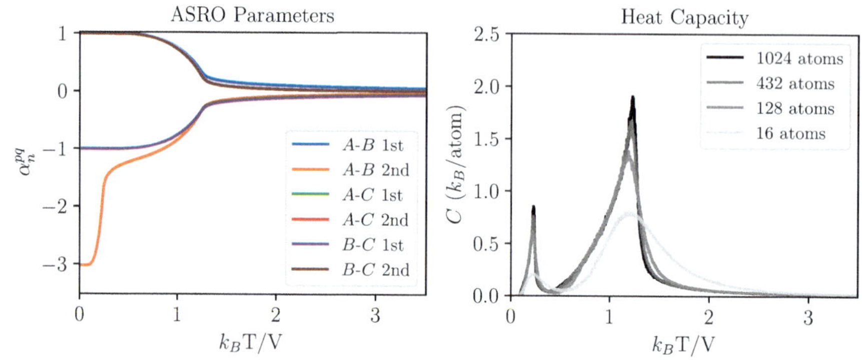


**Fig. 2.9** Visualisation of the Warren-Cowley ASRO parameters (left) and specific heat capacity (right) as a function of temperature for a Monte Carlo simulation of a three component Heusler system on the bcc lattice exhibiting an $L2_1$ ordering. The ASRO parameters are extracted from a simulation on $8 \times 8 \times 8$ cubic bcc unit cells, containing a total of 1024 atoms. The SHC is plotted for a variety of system sizes to demonstrate the convergence with respect to system size

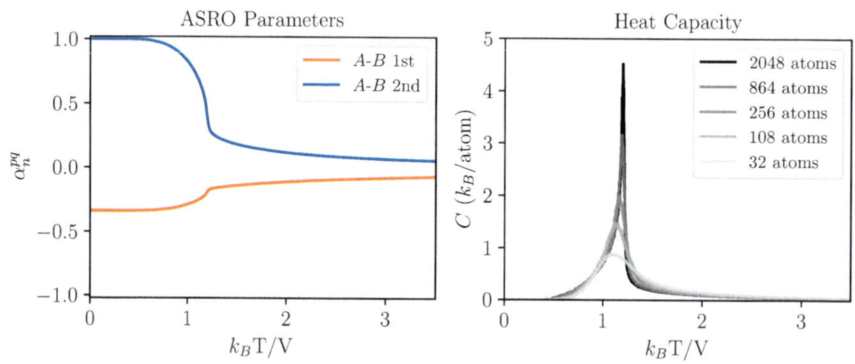


**Fig. 2.10** Visualisation of the Warren-Cowley ASRO parameters (right) and specific heat capacity (left) as a function of temperature for a Monte Carlo simulation of a system on the fcc lattice exhibiting an $L1_0$ ordering. The ASRO parameters are extracted from a simulation on $8 \times 8 \times 8$ cubic fcc unit cells, containing a total of 2048 atoms. The SHC is plotted for a variety of system sizes to demonstrate the convergence with respect to system size

parameters for the simulation were left unchanged, and results were averaged over an ensemble of 32 simulations for each system size.

The results for this simulation are visualised in Fig. 2.10. A quick glance at Fig. 2.2, and an inspection of the Warren-Cowley ASRO parameters as a function of temperature, should convince the reader that an $L1_0$ ordering has been achieved. And, again, the reduced transition temperature compared to the mean-field estimate can be seen. We also observe the convergence of the peak in SHC with respect to system size. The $A1 \rightarrow L1_0$ transition is, strictly, a first-order transition, so we see the height of the peak in the SHC curve, indicative of the phase transition occurring, diverging as the system size increases.

These results from simple, lattice-based Monte Carlo simulations demonstrate how a recovered effective pair interaction facilitates rapid exploration of the phase space of a given multicomponent system. Not only are our estimates of transition temperatures obtained via these simulations considered more reliable compared to the purely mean-field estimates but, as illustrated in the case of the Heusler system, we can learn more about the nature of chemical ordering below any initial disorder-order transition, where the Landau theory no longer applies. In addition, in systems with many constituent elements, such as high-entropy alloys , where the Landau theory alone may not give a complete picture of the phase behaviour, these lattice-based, atomistic simulations should facilitate improved understanding of atomic ordering tendencies.

## 2.11 Summary

In summary, in this chapter, the techniques of statistical physics have been applied to a lattice-based description of a multicomponent alloy. Through use of a Landau-type theory, it has been shown that order-disorder transitions can be estimated directly. In addition, it has been shown that derivatives of an appropriate free energy can be mapped onto a simple, pairwise model for the internal energy of the alloy, which can be studied via, *e.g.*, Monte Carlo simulations, or other appropriate sampling techniques.

The key quantities that it will be necessary to obtain to use this approach to model real systems are: the internal energy for a given configuration, $E_{\text{int}}[\{\xi_{i\alpha}\}]$; the ensemble average of that energy, $\langle E_{\text{int}}[\{c_{i\alpha}\}]\rangle_0$; and the second derivative of that energy with respect to the on-site concentrations, $\tilde{S}^{(2)}_{i\alpha j\alpha'} := \frac{\partial^2 \langle E_{\text{int}}\rangle_0}{\partial \bar{c}_{i\alpha}\partial \bar{c}_{j\alpha'}}\Big|_{\bar{c}_{is}}$. Evaluation of these quantities *ab initio* using electronic structure calculations will be discussed in the following chapter.

For the reader *not* interested in the details of density functional theory, it is possible to assume that these quantities can be evaluated *somehow* from electronic structure calculations, and instead proceed directly to Chap. 4 for the first results.

## References


1. L. Lannelongue, J. Grealey, M. Inouye, Adv. Sci. **8**(12), 2100707 (2021)
2. L. Shenoy, C.D. Woodgate, J.B. Staunton, A.P. Bartók, C.S. Becquart, C. Domain, J.R. Kermode, Phys. Rev. Mater. **8**(3), 033804 (2024)
3. Z. Balogh, G. Schmitz, in *Physical Metallurgy* (Elsevier, 2014), pp. 387–559. https://linkinghub.elsevier.com/retrieve/pii/B9780444537706000058
4. W. Pfeiler (ed.), *Alloy Physics: A Comprehensive Reference*, 1st edn. (Wiley, 2007). https://doi.org/10.1002/9783527614196
5. R.M. Martin, *Electronic Structure: Basic Theory and Practical Methods* (Cambridge University Press, Cambridge, UK, 2004)

6. A.S. Eddington, *The Philosophy of Physical Science: Tarner Lectures (1938)* (Cambridge University Press, Cambridge, 2012)
7. S.N. Khan, J.B. Staunton, G.M. Stocks, Phys. Rev. B **93**(5), 054206 (2016)
8. F. Ducastelle, *Order and Phase Stability in Alloys* (North-Holland; New York, NY, USA, 1991)
9. A.G. Khachaturyan, Progr. Mater. Sci. **22**(1), 1 (1978)
10. A.G. Khachaturyan, *Theory of Structural Transformations in Solids* (Dover Publications, Mineola, N.Y., 2008)
11. M.J. Mehl, D. Hicks, C. Toher, O. Levy, R.M. Hanson, G. Hart, S. Curtarolo, Comput. Mater. Sci. **136**, S1 (2017). https://doi.org/10.1016/j.commatsci.2017.01.017
12. D. Hicks, M.J. Mehl, E. Gossett, C. Toher, O. Levy, R.M. Hanson, G. Hart, S. Curtarolo, Comput. Mater. Sci. **161**, S1 (2019). https://doi.org/10.1016/j.commatsci.2018.10.043
13. D. Hicks, M.J. Mehl, M. Esters, C. Oses, O. Levy, G.L. Hart, C. Toher, S. Curtarolo, Comput. Mater. Sci. **199**, 110450 (2021). https://doi.org/10.1016/j.commatsci.2021.110450
14. L.D. Landau, Zh. Eksp, Teor. Fiz. **7**, 19 (1937)
15. L.D. Landau, Zh. Eksp, Teor. Fiz. **7**, 627 (1937)
16. L.D. Landau, E.M. Lifshitz, *Statistical Physics*, 3rd edn. No. 5 in Course of Theoretical Physics (Elsevier Butterworth-Heinemann, Amsterdam Heidelberg, 1980)
17. R.P. Feynman, *Statistical Mechanics: A Set of Lectures* (Advanced Book Classics (Westview Press, Boulder, Colorado, 1998)
18. P.M. Chaikin, T.C. Lubensky, *Principles of Condensed Matter Physics* (Cambridge University Press, 1995)
19. X. Chen, Q. Wang, Z. Cheng, M. Zhu, H. Zhou, P. Jiang, L. Zhou, Q. Xue, F. Yuan, J. Zhu, X. Wu, E. Ma, Nature **592**(7856), 712 (2021)
20. J.B. Staunton, D.D. Johnson, F.J. Pinski, Phys. Rev. B **50**(3), 1450 (1994)
21. L.J. Santodonato, P.K. Liaw, R.R. Unocic, H. Bei, J.R. Morris, Nat. Commun. **9**(1), 4520 (2018)
22. M.C. Troparevsky, J.R. Morris, P.R. Kent, A.R. Lupini, G.M. Stocks, Phys. Rev. X **5**(1), 011041 (2015)
23. E. Mendive-Tapia, J.B. Staunton, Phys. Rev. Lett. **118**(19), 197202 (2017)
24. E. Mendive Tapia, *Ab initio Theory of Magnetic Ordering: Electronic Origin of Pair- and Multi-Spin Interactions*. Springer Theses (Springer International Publishing, Cham, 2020). http://link.springer.com/10.1007/978-3-030-37238-5
25. W.L. Bragg, E.J. Williams, Proceed. R. Soc. London. Ser. A, Contain. Pap. Math. Phys. Charact. **145**(855), 699 (1934)
26. W.L. Bragg, E.J. Williams, Proceed. R. Soc. London. Ser. A-Math. Phys. Sci. **151**(874), 540 (1935)
27. W. Kohn, Phys. Rev. Lett. **76**(17), 3168 (1996)
28. D.P. Landau, K. Binder, *A Guide to Monte Carlo Simulations in Statistical Physics*, 4th edn. (Cambridge University Press, Cambridge, UK, 2014)
29. K. Binder, J.L. Lebowitz, M.K. Phani, M.H. Kalos, Acta Metallurgica **29**(9), 1655 (1981)
30. M.P. Allen, D.J. Tildesley, *Computer Simulation of Liquids*, 2nd edn. (Oxford University Press, Oxford, United Kingdom, 2017)
31. J.M. Cowley, Phys. Rev. **77**(5), 669 (1950)
32. J.M. Cowley, Phys. Rev. **138**(5A), A1384 (1965)

# Chapter 3
# Electronic Structure *Ab Initio*

**Abstract** We outline the details of a theory underpinning the description of the electronic 'glue' of an alloy, bonding atoms together and driving atomic ordering tendencies. The chapter begins with a brief review of the underlying theory of the electronic structure of condensed matter, and how density functional theory (DFT) makes it possible to (approximately) describe the electronic structure of many systems in a computationally tractable manner. We then proceed to outline the details of the Korringa-Kohn-Rostoker (KKR) formulation of density functional theory, which enables us to review both the coherent potential approximation (CPA) and the disordered local moment (DLM) picture. We then outline the $S^{(2)}$ theory for multicomponent alloys and state the set of coupled equations which must be solved self-consistently to evaluate the atomic short-range order *ab initio*. We also state the reduced form of these equations in the so-called 'band-only' approximation. The chapter concludes by discussing the electronic structure of some well-known metals and alloys, to provide intuition to the reader.

## 3.1 Introduction

Central to modern theoretical condensed matter physics is the many body problem. That is, given a system of interacting particles, one would like first to write down a set of equations describing the system, and then solve to obtain relevant physical quantities. Typically the description used is quantum mechanical, and overwhelmingly the system of interest is one of interacting electrons and nuclei.

Inevitably, to model the physics of alloy formation accurately at the atomic level, one is led to this problem; an accurate description of the electronic 'glue' bonding the atoms together is required. It is these same electrons that are also the origin of other materials quantities of interest, such as structural, electrical, and magnetic properties. In Chap. 2 we saw that, for a given alloy configuration specified by the occupation numbers $\{\xi_{i\alpha}\}$, there were three key quantities of interest. The first was

 
C. D. Woodgate, *Modelling Atomic Arrangements in Multicomponent Alloys*,
Springer Series in Materials Science 346,
https://doi.org/10.1007/978-3-031-62021-8_3

the total energy associated with that configuration,

$$E_{\mathrm{int}}[\{\xi_{\mathrm{i}\alpha}\}]. \tag{3.1}$$

The second was the (mean-field) average of that energy over an ensemble of configurations and the associated average site occupancies, $\{c_{\mathrm{i}\alpha}\}$,

$$\langle E_{\mathrm{int}}\rangle_0[\{c_{\mathrm{i}\alpha}\}]. \tag{3.2}$$

The third quantity was the two-point correlation function, related to the second-derivative of the averaged energy,

$$S^{(2)}_{\mathrm{i}\alpha\mathrm{j}\alpha'} \equiv \left.\frac{\partial^2\langle E_{\mathrm{int}}\rangle_0}{\partial c_{\mathrm{i}\alpha}\partial c_{\mathrm{j}\alpha'}}\right|_{\{c_\alpha\}}. \tag{3.3}$$

The aim of this chapter is to outline a methodology by which these quantities can be evaluated *ab initio*[1] via calculations of a material's electronic structure.

In principle, it is possible to write down a Hamiltonian capturing all the physics of a given system from the view of interacting electrons and nuclei, including relativistic effects if needed. However, for all but the simplest systems the resulting equations are intractable; they cannot be solved analytically because of the complex terms describing interactions between particles, and they cannot be solved numerically because of the large number of variables involved. It is therefore necessary to make approximations of one form or another. This might involve simplifying a Hamiltonian sufficiently that it can be solved analytically, but frequently modern approaches are simply to make approximations to reduce the complexity of the equations describing a system to a level at which they become tractable on the modern computer.

Perhaps the most widespread computational approach used today in the context of materials research is density functional theory (DFT). This builds on the seminal work of Hohenberg and Kohn [1], who demonstrated that all ground state properties of a given system can be obtained from the electron density rather than the full wavefunction. Pioneering work by Kohn and Sham [2] made the further crucial step that instead of solving for the full interacting system, we can reduce the system to a non-interacting auxiliary system in which we model the behaviour of a single electron moving in the effective potential generated by all the other electrons and nuclei in the system. This means the problem is reduced to solving an independent particle equation self-consistently. That is, the effective potential generates the electron density, and the electron density generates the effective potential (up to some numerical tolerance). The method is not without its flaws: the effects of exchange and correlation between electrons have to be captured in some approximate way, and there are many strongly correlated systems which cannot yet be accurately described by

[1] '*Ab initio*' is Latin for 'from the beginning'. The expression 'from first-principles' is also sometimes used. A calculation is said to be *ab initio* if it relies only on fundamental constants and laws of nature with no additional parameters required.

density functional theory. However, for many systems, DFT is capable of obtaining quantities of interest with phenomenal accuracy.

This chapter is laid out in the following way. First, we will outline the theoretical background and key equations describing the many-body problem of interacting electrons and nuclei. We will then outline the basic premises of DFT, which will allow us to reduce the description of a full interacting system of electrons and nuclei to a single-particle description based on the electron density which can be solved self-consistently. We will then describe the technique used in this work to solve the one-electron equations, multiple scattering theory (MST), before going on to discuss the technique used for averaging over configurations, the coherent potential approximation (CPA). These techniques are well suited to describing systems with inherent disorder and also those for which relativistic effects must be taken into account, and will enable numerical evaluation of (3.1) and (3.2). We will also briefly outline the disordered local moment (DLM) picture, which aims to describe the electronic structure of a material as modified by thermally induced magnetic fluctuations. Above the Curie temperature of a system, the DLM picture describes the paramagnetic state, which turns out to be crucial for modelling the physics of alloy formation accurately. We will then briefly outline the scheme derived by Khan, Staunton, and Stocks [3] which provide a means by which to evaluate (3.3). Finally, we will present examples of the electronic structure of some prototypical metals and alloys to furnish the reader with physical intuition, and to assist in the comprehension of results from later chapters.

## 3.2 Theoretical Background

It is worth briefly reviewing the many-body problem associated with a system of interacting electrons and nuclei. The governing equation for a non-relativistic, many-body quantum system containing $N$ particles and with time-independent Hamiltonian is the time-dependent Schrödinger equation,

$$i\hbar \frac{\mathrm{d}}{\mathrm{d}t} \Psi(\mathbf{r}_1, \ldots, \mathbf{r}_N; t) = H\Psi(\mathbf{r}_1, \ldots, \mathbf{r}_N; t), \tag{3.4}$$

where $\Psi(\mathbf{r}_1, \ldots, \mathbf{r}_N; t)$ denotes the full many-body wavefunction. The eigenstates of this equation are solutions of the time-independent Schrödinger equation,

$$H\psi_n(\mathbf{r}_1, \ldots, \mathbf{r}_N) = E_n \psi_n(\mathbf{r}_1, \ldots, \mathbf{r}_N), \tag{3.5}$$

where, when describing fermions, we must also impose the requirement that the full wavefunction be antisymmetric with respect to particle exchange to satisfy the Pauli exclusion principle. The time evolution of an eigenstate is given by

$$\psi_n(\mathbf{r}_1, \ldots, \mathbf{r}_N; t) = \psi_n(\mathbf{r}_1, \ldots, \mathbf{r}_N) e^{-\frac{iE_n t}{\hbar}}. \tag{3.6}$$

A general solution can then be expanded in terms of these eigenstates:

$$\Psi(\mathbf{r}_1, \ldots, \mathbf{r}_N; t) = \sum_n A_n \psi_n(\mathbf{r}_1, \ldots, \mathbf{r}_N) e^{-\frac{iE_n t}{\hbar}}. \tag{3.7}$$

## 3.2.1 The Many-Body Hamiltonian

A basic Hamiltonian for a system of interacting electrons and nuclei can be written as [4]:

$$\begin{aligned} H = &-\frac{\hbar^2}{2m_e}\sum_i \nabla_i^2 - \sum_{i,I}\frac{Z_I e^2}{|\mathbf{r}_i - \mathbf{R}_i|} + \frac{1}{2}\sum_{i \neq j}\frac{e^2}{|\mathbf{r}_i - \mathbf{r}_j|} \\ &- \sum_I \frac{\hbar^2}{2M_I}\nabla_I^2 + \frac{1}{2}\sum_{I \neq J}\frac{Z_I Z_J e^2}{|\mathbf{R}_i - \mathbf{R}_j|}, \end{aligned} \tag{3.8}$$

where the terms capture (in order): the kinetic energy of the electrons; the Coulomb interactions between electrons and nuclei; the Coulomb interactions between the electrons themselves; the kinetic energy of the nuclei; and the Coulomb interactions between the nuclei.

The first observation to make is that one term can be considered ‘small’ compared to the others: the term capturing the kinetic energy of the nuclei, with its factor of $\frac{1}{M_I}$. This is the basis for the ‘adiabatic’ or ‘Born-Oppenheimer’ approximation, which can be expressed physically as the assumption that the timescale of electron motion is so small compared to the timescale of nuclear motion that the nuclei can be assumed to be stationary for the purposes of calculating the eigenstates of the electronic system. In some cases, this approximation is not sufficiently accurate, but even then it can be used as the starting point for a perturbative expansion. Using this approximation, we ‘freeze’ the nuclei in place, setting the term representing the kinetic energy of the nuclei to zero. We are then also able to drop the term capturing the Coulomb interaction between nuclei, as this now only makes a constant contribution—equivalent to changing the ‘zero’ of our energy scale.

To simplify our equations further, we will work in atomic units,[2] that is $\hbar = m_e = e = \frac{4\pi}{\varepsilon_0} = 1$. Our Hamiltonian for the electronic system then takes the form:

$$H = -\frac{1}{2}\sum_i \nabla_i^2 + \sum_i V_{\text{ext}}(\mathbf{r}_i) + \frac{1}{2}\sum_{i \neq j}\frac{1}{|\mathbf{r}_i - \mathbf{r}_j|}, \tag{3.9}$$

where $V_{\text{ext}}$ denotes the external potential associated with the nuclei, $V_{\text{ext}}(\mathbf{r}_i) = -\sum_I \frac{Z_I}{|\mathbf{r}_i - \mathbf{R}_I|}$. It should be noted that other terms can be readily included in this expression if they are pertinent to the problem.

[2] It is a simple procedure to recover quantities of interest in another system of units as a post-processing step.

Whilst this represents a vast simplification of the original problem, it is still the case that for all but the simplest systems this Schrödinger equation cannot be solved for the wavefunction (even numerically) and further approximations are needed.

Up to now, everything that has been said revolves around the idea that the fundamental quantity of interest for a given system is the full electronic wavefunction. In what follows, however, we will demonstrate that to learn about ground state properties of the system, one need only consider the ground state electron density. This is the basis for the construction of the Kohn-Sham auxiliary system on which all modern density-functional theory calculations are based.

## 3.3 The Basics of Density Functional Theory

Density functional theory is ubiquitous in the field of condensed matter physics, and it would be foolish to attempt to describe fully the breadth and depth of the field. There are many excellent textbooks which the reader is invited to consult, such as references [4, 5]. For this exposition, we will therefore only briefly outline the process by which one arrives at the Kohn-Sham equations, outlining key details needed to understand later sections.

Fundamentally, there are two basic premises of DFT. The first is that we can choose to work with the ground state electron density, rather than the ground state wavefunction of the system. This is because the ground state electron density can be shown to determine the full ground-state wavefunction of the system, and the energy of said ground state can be defined as a functional of the ground state electron density. (Hence 'density functional' theory.) The second key premise is that the complex, interacting system can be replaced by a fictitious, non-interacting auxiliary system, described by the Kohn-Sham equations. While the replacement of the complex interacting system with this auxiliary system is, in principle, exact, the functional capturing the effects of electronic exchange and correlation is not known and must be given in some approximate form for practical calculations. The Kohn-Sham equations can be therefore be viewed as providing a means for determining the ground state electron density minimising the energy of the system.

It is possible to develop density functional theory in the relativistic setting from the Dirac equation,[3] and in the non-relativistic setting from the Schrödinger equation outlined above. For the purposes of calculations on some magnetic materials, and in materials containing heavy ions, relativistic effects are important and it is necessary to use the relativistic setting. However, this adds a layer of conceptual difficulty when describing the process by which one arrives at the Kohn-Sham equations. We will therefore outline the derivation in the non-relativistic setting, and then discuss the relativistic case. To further simplify the outline, we will also neglect spin polarisation.

---

[3] In the relativistic setting, one actually works with a 'four-current' density, which arises naturally from the Dirac spinor.

### 3.3.1 The Hohenberg-Kohn Theorems

As outlined above, the key concept behind density functional theory (and the origin of its name) is that *all ground state properties of the system are determined by the ground state electron density*. This in principle greatly simplifies the many-body problem, as it is now sufficient to know the electron density, $n(\mathbf{r})$, rather than the full wavefunction $\Psi(\{\mathbf{r}_i\})$.

That the electron density is a suitable quantity of study was demonstrated by Hohenberg and Kohn, who gave two key results [1]:

**Theorem 3.1** *Given an interacting electron gas moving in an external potential $V_{ext}(\mathbf{r})$, the potential is determined uniquely (up to a constant) by the ground state particle density, $n_0(\mathbf{r})$.*

**Theorem 3.2** *The total energy of the system can be defined as a universal functional of the ground state density, $E = E[n(\mathbf{r})]$ for any potential $V_{ext}(\mathbf{r})$. The ground state energy is the global minimum of this functional, and the global minimiser is exactly the ground state density, $n_0(\mathbf{r})$.*

The consequences of these theorems are profound. The first theorem implies that the Hamiltonian of the system is determined by the ground state density. However, if the Hamiltonian is fully determined, so too are the many-body wave functions for all states, meaning we can write $\Psi = \Psi[n]$. The second results imply that minimisation of the universal functional $E[n]$ determines the energy and electron density of the ground state.

Following (3.9), the total energy functional is written as:

$$\begin{aligned} E[n] &= \langle \Psi[n] | T + E_{\text{int}} | \Psi[n] \rangle + \int \mathrm{d}\mathbf{r}\, n(\mathbf{r}) V_{\text{ext}}(\mathbf{r}) \\ &= F_{\text{HK}}[n] + \int \mathrm{d}\mathbf{r}\, n(\mathbf{r}) V_{\text{ext}}(\mathbf{r}) \end{aligned} \tag{3.10}$$

where it should be emphasised that the Hohenberg-Kohn functional, $F_{\text{HK}}$, is universal and does not depend on the positions of the ions.

The question now is how to make use of this reformulation of the electronic many-body problem in terms of the density. Although in principle $F_{\text{HK}}$ is sufficient to tell us everything we would like to know about the system, the theorems only tell us that this functional exists, not how to go about evaluating it!

### 3.3.2 The Kohn-Sham Equations

The proposal of Kohn and Sham was to use the work of Hohenberg and Kohn to formulate an auxiliary independent-particle problem [2]. The key reason this is useful is that it means that *we can obtain the ground-state electron density by solving an auxiliary non-interacting system, rather than the full interacting system.* The density is now written as

$$n(\mathbf{r}) = \sum_{i=1}^{N} |\psi_i(\mathbf{r})|^2, \tag{3.11}$$

where $N$ is the number of electrons and the $\psi_i$ are solutions to the one-particle Schrödinger equation, to be defined presently. The Hohenberg-Kohn functional is split as

$$F_{\text{HK}}[n] = T_0[n] + E_{\text{Hartree}}[n] + E_{\text{xc}}[n] \tag{3.12}$$

where $T_0$ is now the kinetic energy of the non-interacting electron system, $E_{\text{Hartree}}$ is the Hartree interaction between electrons, and $E_{\text{xc}}$ is the famed exchange-correlation functional, which aims to capture somehow the physics of the full, interacting system by including the effects of electronic exchange and correlation. Explicitly,

$$T_0[n] = -\frac{1}{2}\sum_{i=1}^{N} \int d\mathbf{r}\, \psi_i^*(\mathbf{r})\nabla^2\psi_i(\mathbf{r}), \tag{3.13}$$

and

$$E_{\text{Hartree}}[n] = \frac{1}{2}\int\int d\mathbf{r}\, d\mathbf{r}'\, \frac{n(\mathbf{r})n(\mathbf{r}')}{|\mathbf{r}-\mathbf{r}'|}. \tag{3.14}$$

The exchange-correlation energy aims to account for the difference in energy between interacting and non-interacting systems,

$$E_{\text{xc}}[n] = T[n] - T_0[n] + E_{\text{int}}[n] - E_{\text{Hartree}}[n], \tag{3.15}$$

where $T[n]$ and $E_{\text{int}}[n]$ are the kinetic and internal energies of the full, interacting system. Its form is unknown in general, so in calculations it is approximated. The hope is that its contribution is relatively small, and that the chosen approximation captures it accurately. The choice of form of exchange-correlation functional is one of the main challenges that the modern DFT practitioner faces.

It remains to describe the wave functions $\psi_i$ by deriving the one-particle Schrödinger equation. The Kohn-Sham energy can be written as

$$E_{\text{KS}}[n] = T_0[n] + E_{\text{Hartree}}[n] + E_{\text{xc}}[n] + \int d\mathbf{r}\, V_{\text{ext}}(\mathbf{r})n(\mathbf{r}). \tag{3.16}$$

Following the variational principle, we seek to minimise this functional with respect to the density $n$. Given that the density is to be constructed from the one-particle orbitals as in (3.11), we apply the chain rule and obtain [4]

$$\frac{\delta E_{\text{KS}}}{\delta\psi_i^*} = \frac{\delta T_0}{\delta\psi_i^*} + \left[\frac{\delta E_{\text{Hartree}}}{\delta n} + \frac{\delta E_{\text{xc}}}{\delta n} + \frac{\delta E_{\text{ext}}}{\delta n}\right]\frac{\delta n}{\delta\psi_i^*} = 0. \tag{3.17}$$

Using the definitions of $T_0$ and $n$, we can show that

$$\frac{\delta T_0}{\delta \psi_i^*} = -\frac{1}{2}\nabla^2 \psi_i, \tag{3.18}$$

and

$$\frac{\delta n}{\delta \psi_i^*} = \psi_i. \tag{3.19}$$

Then we write

$$\begin{aligned} V_{\text{KS}} &= V_{\text{ext}} + \frac{\delta E_{\text{Hartree}}}{\delta n} + \frac{\delta E_{\text{xc}}}{\delta n} \\ &= V_{\text{ext}} + V_{\text{Hartree}} + V_{\text{xc}}, \end{aligned} \tag{3.20}$$

defining our effective potential. We constrain our wavefunctions to be orthonormal, $\langle \psi_i | \psi_j | = \delta_{ij}$ using a Lagrange multiplier, $\varepsilon_i$, to arrive at the famous Kohn-Sham equations

$$[H_{\text{KS}} - \varepsilon_i]\,\psi_i = 0, \tag{3.21}$$

where $H_{\text{KS}}(\mathbf{r}) = -\frac{1}{2}\nabla^2 + V_{\text{KS}}(\mathbf{r})$. This one-particle Schrödinger equation is the basis for practically all modern DFT calculations.

It is important to note that the effective potential in these equations is dependent on the density itself. The equations therefore have to be solved self-consistently to obtain the electronic density. Generally, this means starting with an initial 'guess' at the electron density of the system. Using this density, the effective potential is defined and Kohn-Sham equations can be solved to determine a new electron density. Self-consistency is achieved when the input electron density (determining the effective potential) is equal to the output electron density (determined *by* the effective potential) up to some numerical tolerance.

### 3.3.3 The Local Density Approximation

So far we have not specified the form of $E_{\text{xc}}$, the exchange-correlation energy. It is not known exactly, and the search for 'good' exchange-correlation functionals continues to this day [4, 5]. There are many such functionals, and the choice of functional can have significant impact on numerical results. It is also true that different functionals are suited to different systems. For example, we intuitively understand that the effects of exchange and correlation are likely to be different in a close-packed, periodic, metallic system compared to an isolated organic molecule!

Perdew and Schmidt proposed the idea of a 'Jacob's Ladder' of functionals; the idea that increasingly advanced functionals include more and more complex electronic effects, enable us to climb a ladder of increasing accuracy towards the 'heaven' of true chemical accuracy [6]. For the purposes of calculations on metals, however, the lowest rung of this ladder is usually sufficient. This is a family of exchange-correlation functionals making use of the local density approximation (LDA), which was proposed by Kohn and Sham in their original paper [2]. The LDA also happens to be a commonly used numerical scheme because it is both physically intuitive and easily implemented.

The key idea behind the LDA is that the exchange correlation energy is assumed to only depend *locally* on the computed electron density itself, not its derivatives and not on the orbitals themselves. In general, in the LDA, the exchange-correlation functional takes the form:

$$E_{\mathrm{xc}}[n] = \int \mathrm{d}\mathbf{r}\, n(\mathbf{r})\varepsilon_{\mathrm{xc}}\left[n(\mathbf{r})\right], \tag{3.22}$$

where $\varepsilon_{\mathrm{xc}}$ is a scalar function taking only the value of the electron density at a specific point, $\mathbf{r}$, as its argument.

Although this seems a crude approximation (and, in fact, it is) it turns out that the LDA satisfies an important sum rule concerning the so-called exchange-correlation hole, describing the effect of inter-electronic repulsions and the consequence that an electron being present at one point, $\mathbf{r}$, reduces the probability of finding one at another point, $\mathbf{r}'$. It can be shown that the exchange-correlation hole must contain exactly one electron, and that the LDA fulfills this important requirement [7].

LDA functionals are typically based on the homogeneous electron gas, for which the exchange energy is known exactly [8]. The correlation term is obtained by parameterising data obtained by numerically solving a many-body Schrödinger equation [9]. One widely used LDA functional, for example, is that of Perdew and Wang [10].

## 3.4 Including Spin, Magnetism, and Relativity in Density Functional Theory

The treatment given above neglected magnetism to elucidate some of the key aspects of density functional theory. However, we will now turn our attention to the inclusion of magnetism, *i.e.* spin polarisation, in the formalism. Spin arises naturally from relativistic quantum mechanics [11], where one seeks to formulate a wave equation akin to the Schrödinger equation but that is Lorentz-invariant. This naturally leads to the Dirac equation for a free electron. The Dirac equation is written

$$\begin{aligned} H_{\mathrm{D}}\Psi_i(\mathbf{r}) &= \left[c\boldsymbol{\alpha}\cdot\mathbf{p} + \beta m_e c^2\right]\Psi_i(\mathbf{r}) \\ &= \varepsilon_i\Psi_i(\mathbf{r}), \end{aligned} \tag{3.23}$$

where $c$ is the speed of light, $\mathbf{p} = -i\hbar\nabla$ is the momentum operator, and $\beta$ is the $4\times 4$ matrix defined as

$$\beta = \begin{pmatrix} I_2 & 0_{2\times 2} \\ 0_{2\times 2} & -I_2 \end{pmatrix}, \tag{3.24}$$

where $I_2$ and $0_{2\times2}$ are the $2 \times 2$ identity matrix and zero matrix, respectively. The vector $\boldsymbol{\alpha} = (\alpha_x, \alpha_y, \alpha_z)$ is defined as having components

$$\alpha_i = \begin{pmatrix} 0_{2\times2} & \sigma_i \\ \sigma_i & 0_{2\times2} \end{pmatrix}, \tag{3.25}$$

where $\{\sigma_i\}$ are the Pauli matrices,

$$\sigma_x = \begin{pmatrix} 0 & 1 \\ 1 & 0 \end{pmatrix}, \quad \sigma_y = \begin{pmatrix} 0 & -i \\ i & 0 \end{pmatrix}, \quad \text{and} \quad \sigma_z = \begin{pmatrix} 1 & 0 \\ 0 & -1 \end{pmatrix}. \tag{3.26}$$

The solution of the Dirac equation, (3.23), is the well-known Dirac spinor, $\Psi_i(\mathbf{r})$, which has four components. Dirac's theory of the electron naturally gives rise to the fundamental quantum mechanical property of the electron spin, as well as to phenomenon such as spin-orbit coupling, which can only be included perturbatively in the non-relativistic setting [12].

It is perfectly possible (although lengthy) to derive a relativistic version of density functional theory, starting from quantum electrodynamics and following similar arguments to those of the non-relativistic case. Relativistic effects turn out to be important in the description of the electronic structure of systems containing heavy atoms, and also in describing phenomena such as magnetocrystalline anisotropy, where spin-orbit coupling connects magnetic moments to the crystalline lattice of a material. One approximation that can be made to simplify the equations which need to be solved and, therefore, to reduce the associated computational cost is the so-called scalar-relativistic approximation, where the Dirac equation is reduced by omitting the spin-orbit interaction. The scalar-relativistic approximation results in an equation of similar structure to the Schrödinger equation, which is sometimes referred to as the 'scalar-relativistic wave equation' [12].

In spin-polarised DFT, in either its non-relativistic or scalar-relativistic versions, the electron density $n(\mathbf{r})$ is supplemented by the magnetisation density $\mathbf{M}(\mathbf{r})$. Equivalently, the $2 \times 2$ spin density matrix, $n^{\alpha\alpha'}$ can be used. The electron density is given by

$$n(\mathbf{r}) = \mathrm{Tr}\left(n^{\alpha\alpha'}(\mathbf{r})\right), \tag{3.27}$$

while the magnetisation density, in general, is given by

$$\mathbf{M}(\mathbf{r}) = \sum_{\alpha\alpha'} \boldsymbol{\sigma}^{\alpha\alpha'} n^{\alpha\alpha'}(\mathbf{r}) \tag{3.28}$$

where $\boldsymbol{\sigma} = (\sigma_x, \sigma_y, \sigma_z)$ is the vector of Pauli matrices. This describes the general case where there is no common global magnetisation axis for all atoms in the system, *i.e.* spins are non-collinear. A simplification of this can be applied to systems where there *is* a common axis of global magnetisation (describing ferromagnets, antiferromagnets, and ferrimagnets), where a global $z$ axis can be chosen, the magnetisation vector can be replaced with its magnitude $\mathbf{M}(\mathbf{r}) \to |\mathbf{M}(\mathbf{r})|$, and the spin-density matrix can be replaced by a spin-up and spin-down densities, $n^{\uparrow}(\mathbf{r})$ and $n^{\downarrow}(\mathbf{r})$. Calculations performed within this simplification are referred to as collinear. In both collinear and non-collinear cases, there are generalised exchange-correlation functionals available for practical calculations.

## 3.5 Multiple Scattering Theory: A Route to Self-Consistent Calculations

Regardless of the treatment of spin polarisation, the inclusion of relativistic effects, and the choice of exchange-correlation functional, for a given system we require a self-consistent solution of a set of (Schrödinger-like) Kohn-Sham equations,

$$\left[-\frac{1}{2}\nabla^2 + V_{\text{eff}} - \varepsilon_i\right]\psi_i = 0. \tag{3.29}$$

It is then necessary to find the self-consistent electron density which satisfies the equation. This will then be the ground state electron density of the system of interest. The procedure followed is to start with some initial 'guess' at the ground state electron density (which determines the exchange-correlation energy), then solve for the independent particle solutions, before reconstructing a new electron density. When the new electron density is within some numerical tolerance of the original, self-consistency has been reached.

The form of the Kohn-Sham equations (particularly in the non-relativistic setting) naturally lend themselves to a description of solutions in terms of a (hopefully small) set of basis functions. Two common choices of basis function are plane waves (for periodic solids) and atomic orbitals (for isolated molecules). However, this approach has a major drawback; there is no meaningful way to find the average properties of a disordered system, as the geometry of the set of ions being considered is inextricably linked to the nature of those ions themselves.

An alternative approach which separates the geometry of the system from the nature of ionic sites is the Korringa-Kohn-Rostoker (KKR) formalism, also referred to as multiple scattering theory (MST) [13–15]. This technique is based on formulating solutions in terms of Green's functions [16]. It proves conceptually more challenging but offers considerable advantages when considering disordered systems. The picture to have in mind is that of a single electron propagating through a periodic system and being 'scattered' by individual ions as it does so.

Taking the viewpoint that we imagine an itinerant electron propagating through the lattice and being scattered by the ions at positions $\{\mathbf{R}_{\text{i}}\}$, we first decompose the effective potential in the Kohn-Sham equations into a sum over separate potentials localised around the ions,

$$V_{\text{eff}}(\mathbf{r}) = \sum_{\text{i}} V_{\text{i}}(\mathbf{r}_{\text{i}}) \tag{3.30}$$

where $\mathbf{r}_\mathrm{i} = \mathbf{r} - \mathbf{R}_\mathrm{i}$, and the individual potentials $V_\mathrm{i}$ are localised in space (*i.e.* there is some radius $R$ for which $V_\mathrm{i}(\mathbf{r}) = 0$ when $|\mathbf{r}| > R$). There are typically three approaches used to describe these spatially bounded potentials. The first (and simplest) is the so-called muffin-tin (MT) approximation where there is no overlap between the potentials on different lattice sites; the potential is spherically symmetric within the muffin-tin radius and flat outside of it. Also widely used is the Atomic Sphere Approximation (ASA), which very similar to the MT approximation, but allows the potentials to overlap slightly because the radius of the region where the potential is non-zero is set to be a sphere with volume equal to that of the Wigner-Seitz cell, determining the Wigner-Seitz radius. Finally, a number of modern KKR codes make use of space-filling potentials which have no spherical symmetry. The potential is then defined on polyhedra which tile 3-space. These are referred to as 'full-potential' calculations and have the advantage that they can successfully handle a wider range of systems than either the muffin-tin or atomic-sphere approximations. Full-potential calculations also turn out to be important when evaluating forces on ions. However, for the close-packed, itinerant electron systems of this work the MTA and ASA are both computationally cheap and satisfactorily accurate, so we will restrict our discussion to these approximations.

We will now briefly introduce MST in the conventional manner, first by discussing the case of scattering from a single, isolated scatterer, then building on this to describe the case of multiple scatters. Finally, we demonstrate how the electronic density can be arrived at directly from the Green's function of a system. This lays the foundation for the description of chemical disorder in Sect. 3.6, the description of magnetic disorder in Sect. 3.7, and the $S^{(2)}$ theory derived within this framework as outlined in Sect. 3.8. The author notes that there are many books which cover this topic in far greater detail than in this work. The reader is invited to consult references [13–16].

### 3.5.1 Single-Potential Scattering

The fundamental idea of MST is that the total Green's function of the system can be assembled from independent, local solutions of the Schrödinger-like equation centred around each lattice site i. The procedure therefore begins by numerically solving this local Schrödinger equation at site i,

$$H_\mathrm{i} \left|\psi_\mathrm{i}\right\rangle = \left[-\frac{1}{2}\nabla^2 + V_\mathrm{i}\right]\left|\psi_\mathrm{i}\right\rangle = E\left|\psi_\mathrm{i}\right\rangle, \tag{3.31}$$

for two linearly independent solutions $\phi_{\mathrm{i}L}(E;\mathbf{r})$ and $J_{\mathrm{i}L}(E;\mathbf{r})$ for each composite angular momentum index $L = (l, m)$. These two functions are determined by the imposed boundary conditions

$$\lim_{|\mathbf{r}|\to 0} \phi_{iL}(E;r) \to j_l(\sqrt{E}r)Y_{lm}(\mathbf{r})$$

$$\lim_{\mathbf{r}\notin V_i} J_{iL}(E;r) \to j_l(\sqrt{E}r)Y_{lm}(\mathbf{r}),$$

where $j_l$ are the spherical Bessel functions and $Y_{lm}$ are the spherical harmonics. It is helpful to note that a plane wave, $e^{i\mathbf{k}\cdot\mathbf{r}}$ can be expanded in terms of spherical harmonics via

$$e^{i\mathbf{k}\cdot\mathbf{r}} = 4\pi \sum_L i^l Y_L^*(\mathbf{k})Y_L(\mathbf{r})j_l(k\mathbf{r}), \tag{3.32}$$

which is referred to as the 'partial wave expansion' of the plane wave.

It is common to transform the Jost function $\phi_{iL}$ to a new function $Z_{iL}$ for ease of later manipulations. $Z_{iL}$ is defined as

$$Z_{iL}(E;\mathbf{r}) = \sum_{L'} \phi_{iL'}(E;\mathbf{r})(\alpha_i t_i^{-1})_{L'L}, \tag{3.33}$$

where the matrices $\alpha_{i;LL'}(E)$ and, $t_{i;LL'}(E)$, referred to as the $\alpha$-matrix and the $t$-matrix, will be defined below. It is also conventional in the multiple scattering literature to define the 'regular scattering solution', $R_{iL}$, which is defined as

$$R_{iL}(E;\mathbf{r}) = j_l(\sqrt{E}r)Y_{lm}(r) + \int d\mathbf{r}' G_0(E,\mathbf{r},\mathbf{r}')V_i(\mathbf{r}')R_{iL}(E;\mathbf{r}'), \tag{3.34}$$

where the free-particle Green function, $G_0$—describing the propagation of a free electron with no imposed potential—is defined as

$$G_0(E;\mathbf{r}'\mathbf{r}') = \langle\mathbf{r}|(E - H_0)^{-1}|\mathbf{r}'\rangle. \tag{3.35}$$

The $\alpha$-matrix is then defined by the limiting behaviour of the regular scattering solution by

$$\lim_{|\mathbf{r}|\to 0} R_{iL}(E;r) = \sum_{L'} j_{l'}(\sqrt{E}r)Y_{L'}(r)\alpha_{i;L'L}. \tag{3.36}$$

This matrix is used in the definition of Lloyd's formula, defined later, which is an expression which can be used to obtain the integrated electronic density of states, $N(E)$.

From the above solutions of the single-site Schrödinger equation, the $t$-matrix $\underline{t}_i$ describing the scattering for potential $V_i$ is constructed, which relates incoming and outgoing electronic states scattered by the potential at site i. In the angular momentum representation, *i.e.* in a basis of $j_l(\sqrt{er})Y_{lm}(\mathbf{r})$, for a general potential $V_i$, this is defined as

$$t_{i;LL'}(E) = \int d\mathbf{r}\, j_l(\sqrt{E}r)Y_{lm}(\mathbf{r})V_i(\mathbf{r})R_{iL'}(E;\mathbf{r}). \tag{3.37}$$

For a spherically symmetric potential, as in the case of either the MTA or ASA, this matrix is diagonal in $l$ and reduces to

$$t_{\text{i};LL'}(E) = -\delta_{ll'} e^{i\delta_{\text{i}l}(E)} \sin \delta_{\text{i}l}(E)/\sqrt{E}, \tag{3.38}$$

where $\delta_{\text{i}l}(E)$ are the scattering phase shifts corresponding to site i.[4] Note that, in (3.38), $\delta_{ll'}$ denotes the Kronecker delta, while $\delta_{\text{i}l}(E)$ are the phase shifts. This somewhat confusing convention is used elsewhere in the KKR literature and we retain it here for consistency with those other works.

We note that, in all of the above expressions, in practical calculations the sum over angular momentum indices $(l, m)$ is truncated at some finite value. Within the MTA and ASA it is common (and often satisfactorily accurate) to take $l_{\text{max}} = 3$, although in full-potential codes and with modern computing resources it is common to make a much larger expansion.

### 3.5.2 *Scattering Paths and the Multi-site Solution*

Having obtained the scattering matrix for each site i, $\underline{t}$, these matrices are combined to form the full $t$-matrix of the entire system, $\underline{\underline{t}}$, by concatenating them along the diagonal of a super-matrix, with combined row/column index (i, $L$). (The double underline indicates a super-matrix with combined site and angular momentum index.) This super-matrix is defined as

$$t_{\text{i}L;\text{j}L'}(E) = \delta_{\text{ij}} t_{\text{i};LL'}(E). \tag{3.39}$$

From this super-matrix a key quantity of MST is constructed, known as the 'scattering path operator', denoted $\underline{\underline{\tau}}$ and defined via

$$\underline{\underline{\tau}} = \left[\underline{\underline{t}}^{-1} - \underline{\underline{G}}_0\right]^{-1}, \tag{3.40}$$

where the so-called structure constants, $\underline{\underline{G}}_0$ are independent of the potentials $V_{\text{i}}$ and are determined solely by the crystal lattice and basis, $\mathbf{R}_{\text{i}} = \mathbf{R}_i + \mathbf{h}_a$ [15]. The scattering path operator (SPO) is the analogue of the $t$-matrix (for a single potential) for the entire system, connecting an incoming electronic state at site i with an outgoing electronic state at site j.

Finally, from the scattering path operator, the Green function of the entire system can be constructed via [15]

[4] Equations 3.37 and 3.38 represent a significant portion of the challenge associated with full-potential MST; the off-diagonal elements introduce additional complexity in finding a self-consistent solution in a numerically stable manner.

$$G(E;\mathbf{r},\mathbf{r}') = \sum_{LL'} Z_{\mathrm{i}L}(E;\mathbf{r})\tau_{\mathrm{i}L;\mathrm{j}L'}(E)Z_{\mathrm{j}L'}(E;\mathbf{r}') - \delta_{\mathrm{ij}}\sum_{L} Z_{\mathrm{i}L}(E;\mathbf{r}_<)J_{\mathrm{i}L}(E;\mathbf{r}_>), \tag{3.41}$$

where $\mathbf{r} \in V_{\mathrm{i}}$, $\mathbf{r}' \in V_{\mathrm{j}}$, $\mathbf{r}_< = \min(\mathbf{r} - \mathbf{R}_{\mathrm{i}}, \mathbf{r}' - \mathbf{R}_{\mathrm{i}})$, and $\mathbf{r}_< = \min(\mathbf{r} - \mathbf{R}_{\mathrm{i}}, \mathbf{r}' - \mathbf{R}_{\mathrm{i}})$. From the full Green function of the system, the charge density,

$$\rho(\mathbf{r}) = -\frac{1}{\pi}\lim_{\epsilon\to 0}\mathrm{Im}\int f(E-\mu)G(E+i\epsilon;\mathbf{r},\mathbf{r})\mathrm{d}E, \tag{3.42}$$

the density of states,

$$g(E) = -\frac{1}{\pi}\lim_{\epsilon\to 0}\mathrm{Im}\int f(E-\mu)G(E+i\epsilon;\mathbf{r},\mathbf{r})\mathrm{d}\mathbf{r}, \tag{3.43}$$

and the total energy of the system can be obtained. Note that within the MST literature it is generally preferred to work with the charge density, $\rho(\mathbf{r})$, in general related to the electron density by[5] $\rho(\mathbf{r}) = en(\mathbf{r})$. Note also that $f(E-\mu)$ is the conventional Fermi-Dirac distribution, where the chemical potential, $\mu$, ensures overall charge neutrality. In a practical calculation, the integrals of (3.42) and (3.43) are evaluated (typically using a carefully chosen contour in the complex plane) numerically for some finite set of energies, $\{E_m\}$.

### 3.5.3 *Self-Consistent Calculations*

With the Green function introducing an additional level of abstraction to the problem, it can be difficult to see a route by which a self-consistent calculation for a given set of ions can be performed. The procedure is as follows.

Given an initial set of starting potentials $V_{\mathrm{i}}^{\mathrm{in}}$, the electron density $\rho(\mathbf{r})$ is determined via the procedure outlined above. To determine a new potential from this density (within the LDA), the electron density is used to construct a new potential $V_{\mathrm{i}}^{\mathrm{Hartree}}(\mathbf{r})$ by solving the Poisson equation

$$\nabla^2 V_{\mathrm{Hart.}}(\mathbf{r}) = -e\rho(\mathbf{r}), \tag{3.44}$$

and a new exchange-correlation potential, $V_{\mathrm{i}}^{\mathrm{xc}}(\mathbf{r})$. These two potentials are then combined to form the output potential,

$$V_{\mathrm{i}}^{\mathrm{out}}(\mathbf{r}) := V_{\mathrm{i}}^{\mathrm{Hart.}}(\mathbf{r}) + V_{\mathrm{i}}^{\mathrm{xc.}}(\mathbf{r}). \tag{3.45}$$

This procedure can then be repeated (with appropriate mixing of potentials) until $V_{\mathrm{i}}^{\mathrm{out}} = V_{\mathrm{i}}^{\mathrm{in}}$ up to some numerical tolerance. The above procedure therefore outlines how to evaluate the internal energy of the alloy specified in (3.1).

---

[5] In atomic units these two variables are equivalent.

## 3.6 The Coherent Potential Approximation: Averaging Over Disorder

The next quantity of interest is the ensemble average of the internal energy for a given set of averaged site occupancies, $\langle E_{\mathrm{int}}\rangle_0[\{c_{\mathrm{i}\alpha}\}]$. The framework within which this is addressed in this work is referred to as the coherent potential approximation (CPA) [17–19], which is a well-known and widely-used technique for averaging over disordered systems within the framework of MST. This is where the power of the $t$-matrix, SPO, and Green's function become apparent; the abstraction means that averaging over disorder can be done in a more physically meaningful way than if we were working directly with the Kohn-Sham wavefunctions $\{\psi_i\}$. Any disordered system has broken translational symmetry, which means that the Bloch symmetry of a conventional bandstructure calculation is destroyed. The CPA captures this broken translational symmetry in a way that is not possible within conventional bandstructure calculations.[6]

Within the CPA, we seek to describe the average electronic structure of a system with site-wise concentrations $\{c_{\mathrm{i}\alpha}\}$. The single potential at site i, $V_{\mathrm{i}}$, is replaced with a set of single-site potentials, $\{V_{\mathrm{i}\alpha}\}$, associated with each chemical species $\alpha$. From each of these potentials a $t$-matrix can be constructed, $t_{\mathrm{i}\alpha;LL'}$. The CPA seeks an optimal mean-field description of the average scattering properties of this set of $t$-matrices for each site. This means searching for a medium of scatterers $t^c_{\mathrm{i};L,L'}$, where $c$ means CPA, which somehow capture the average electronic structure of the disordered system. In this way the CPA can be thought of as describing an effective medium representing the disordered system. This optimal mean-field medium has an associated SPO

$$\underline{\underline{\tau}}^c = \left[ (\underline{\underline{t}}^c)^{-1} - \underline{\underline{G}}_0 \right]^{-1}, \tag{3.46}$$

which can be used in the same manner as the SPO for an ordered system to obtain quantities of interest. It remains to discuss how the matrices $t^c_{\mathrm{i};L,L'}$ are constructed.

The CPA condition is arrived at by considering an impurity atom $\alpha$ embedded in the disordered, effective medium at site $\mathrm{i}_0$. (This impurity has potential $V_{\mathrm{i}_0\alpha}$.) At this site, the CPA $t$-matrix is replaced by the $t$-matrix for the impurity, resulting in the full $t$-matrix of the system taking the form

$$t^{\mathrm{i}_0\alpha}_{\mathrm{i}L;\mathrm{j}L'} = \delta_{\mathrm{ij}}\delta_{\mathrm{ii}_0} t_{\mathrm{i}\alpha;LL'} + \delta_{\mathrm{ij}}(1-\delta_{\mathrm{ii}_0}) t^c_{\mathrm{i};LL'}. \tag{3.47}$$

An illustration of the scheme is given in Fig. 3.1. The associated SPO can be constructed as

$$\underline{\underline{\tau}}^{\mathrm{i}_0\alpha} = \left[ (\underline{\underline{t}}^{\mathrm{i}_0\alpha})^{-1} - \underline{\underline{G}}_0 \right]^{-1}, \tag{3.48}$$

---

[6] It should be stressed, however, that when an average is taken over a suitably chosen supercell (or set of supercells), physical properties can be calculated very reliably. One means by which to generate appropriate supercells is by use of special quasirandom structures (SQS's) [20].

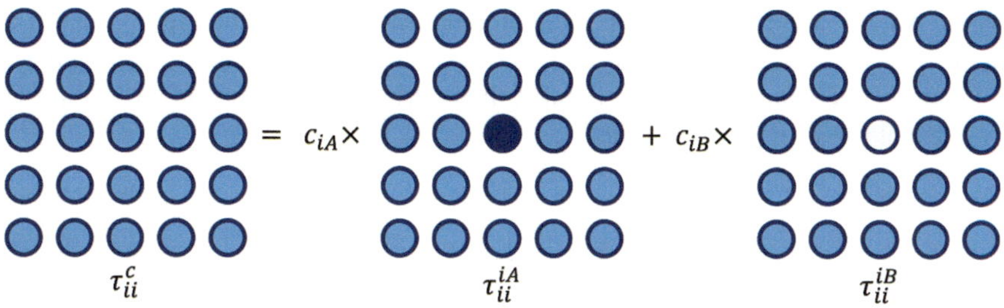

**Fig. 3.1** Illustration of the CPA condition for a 50:50, 2D, binary alloy. Dark blue sites denote those fully occupied by species $A$, white those occupied by species $B$, and pale blue those with occupancy $A_{0.5}B_{0.5}$

from which an associated Green's function, $G^{\mathrm{i}_0\alpha}(E, r, r')$, can be obtained. To fix the reference medium $t^c_{\mathrm{i};L,L'}$, the CPA condition is defined as

$$\tau^c_{\mathrm{ii}} = \sum_\alpha c_{\mathrm{i}\alpha} \tau^{\mathrm{i}\alpha}_{\mathrm{ii}}. \tag{3.49}$$

In words, this condition can be described as the statement that an SPO averaged over impurities at a given site restores the mean-field SPO.

From the above, we see that, given a set of potentials, $\{V_{\mathrm{i}\alpha}\}$, we can construct a set of $t$-matrices and associated SPO $\underline{\underline{\tau}}^{\mathrm{i}\alpha}$. We can therefore define site-dependent charge densities $\rho_{\mathrm{i}\alpha}(\mathbf{r})$ and density of states $g_{\mathrm{i}\alpha}(E)$ from the Green's function $G = G^{\mathrm{i}\alpha}$. It remains to specify how to determine $V_{\mathrm{i}\alpha,\mathrm{out}}(r)$ for use in the self-consistent portion of the calculation. This has been considered in detail by Johnson *et al.* [21] and, within the LDA, can be written as

$$\begin{aligned} V_{\mathrm{i}\alpha}(\mathbf{r}) =& V_{\mathrm{xc}}(\rho_{\mathrm{i}\alpha}(\mathbf{r})) + e^2 \int_{V_\mathrm{i}} \mathrm{d}\mathbf{r}' \frac{\rho_{\mathrm{i}\alpha}(\mathbf{r}') - Z_\alpha \delta(\mathbf{r}' - \mathbf{R}_\mathrm{i})}{|\mathbf{r} - \mathbf{r}'|} \\ &+ e^2 \sum_{\mathrm{j}\neq\mathrm{i}} \int_{V_\mathrm{j}} \mathrm{d}\mathbf{r}' \frac{\overline{\rho}_\mathrm{j}(\mathbf{r}') - \bar{Z}_\mathrm{j}\delta(\mathbf{r}' - \mathbf{R}_\mathrm{j})}{|\mathbf{r} + \mathbf{R}_\mathrm{i} - \mathbf{R}_\mathrm{j} - \mathbf{r}'|}. \end{aligned} \tag{3.50}$$

In this expression, $Z_\alpha$ is the atomic number of atom $\alpha$, while $\overline{\rho}_\mathrm{i}(\mathbf{r}) = \sum_\alpha c_{\mathrm{i}\alpha}\rho_{\mathrm{i}\alpha}(\mathbf{r})$ and $\bar{Z}_\mathrm{i} = \sum_\alpha c_{\mathrm{i}\alpha} Z_\alpha$ are site averages. The second and third terms represent the intra- and inter-site Coulomb interactions respectively. Using this prescription it is possible to take $V^{\mathrm{out}}_{\mathrm{i}\alpha}(\mathbf{r}) \to V^{\mathrm{in}}_{\mathrm{i}\alpha}(\mathbf{r})$ until self-consistency is achieved. Note that this represents an *additional* CPA self-consistency requirement on top of that for the charge density.

Within the CPA, the (average) integrated DoS $\langle N(E)\rangle_0$ is approximated using Lloyd's formula [22], writing

$$N_c(E) = N_0(E) + \frac{1}{\pi}\mathrm{Im}\left\{ \ln ||\tau^c|| + \sum_{\mathrm{i}\mu} c_{\mathrm{i}\mu}(\ln ||\alpha_{\mathrm{i}\mu} t^{-1}_{\mathrm{i}\mu}|| - \ln ||\bar{D}^{-1}_{\mathrm{i}\mu}||)\right\} \tag{3.51}$$

where $N_0(E)$ is the free-electron integrated density of states and the $\alpha_{\mathrm{i}\mu}$ matrix as defined earlier. The determinant $||\tau^c||$ is over indices (i, $L$) while remaining determinants are over indices $L$ only. Finally, the matrix $\bar{D}_{\mathrm{i}\mu}$ comes from the definitions of relevant CPA-related quantities,

$$\Delta_{\mathrm{i}\alpha} = t_{\mathrm{i}\alpha}^{-1} - (t_{\mathrm{i}}^c)^{-1}, \tag{3.52}$$

$$D_{\mathrm{i}\alpha} = [1 + \tau_{\mathrm{ii}}^c \Delta_{\mathrm{i}\alpha}]^{-1} = \Delta_{\mathrm{i}\alpha}^{-1} X_{\mathrm{i}\alpha}, \tag{3.53}$$

$$\bar{D}_{\mathrm{i}\alpha} = [1 + \Delta_{\mathrm{i}\alpha} \tau_{\mathrm{ii}}^c]^{-1} = X_{\mathrm{i}\alpha} \Delta_{\mathrm{i}\alpha}^{-1}. \tag{3.54}$$

Notably, the Lloyd formula obeys a variational property $\delta N_c / \delta (t_{\mathrm{i}}^c)^{-1} = 0$ when varying mean-field medium $\{t_{\mathrm{i}}^c\}$ away from the CPA solution while holding potentials $\{V_{\mathrm{i}\alpha}(\mathbf{r})\}$ fixed.

Finally, from the self-consistent single-site potentials, $\{V_{\mathrm{i}\alpha}\}$, and associated charge densities, $\{\rho_{\mathrm{i}\alpha}\}$, we can now evaluate an expression for the total energy of the alloy within the CPA, $\Omega_c$. The derivation of this has been considered in detail in [21], and the energy is given by

$$\begin{aligned}
\Omega_c = &- \int \mathrm{d}E\, f(E-\mu) N_c(E) - \sum_{\mathrm{i}\alpha} c_{\mathrm{i}\alpha} \int_{V_{\mathrm{i}}} \mathrm{d}\mathbf{r} \rho_{\mathrm{i}\alpha}(\mathbf{r}) V_{\mathrm{i}\alpha}(\mathbf{r}) \\
&+ \Bigg\{ \sum_{\mathrm{i}\alpha} c_{\mathrm{i}\alpha} \int_{V_{\mathrm{i}}} \mathrm{d}\mathbf{r}\, \rho_{\mathrm{i}\alpha}(\mathbf{r}) \epsilon_{\mathrm{xc}}(\rho_{\mathrm{i}\alpha}(\mathbf{r})) \\
&+ \frac{e^2}{2} \sum_{\mathrm{i}\alpha} c_{\mathrm{i}\alpha} \int_{V_{\mathrm{i}}} \mathrm{d}\mathbf{r} \int_{V_{\mathrm{i}}} \mathrm{d}\mathbf{r}' \frac{1}{|\mathbf{r}-\mathbf{r}'|} \left[ \rho_{\mathrm{i}\alpha}(\mathbf{r}) \rho_{\mathrm{i}\alpha}(\mathbf{r}') - 2 Z_\alpha \delta(\mathbf{r}) \rho_{\mathrm{i}\alpha}(\mathbf{r}') \right] \\
&+ \frac{e^2}{2} \sum_{\mathrm{i} \neq \mathrm{j}} \sum_{\alpha\beta} c_{\mathrm{i}\alpha} c_{\mathrm{j}\beta} \int_{V_{\mathrm{i}}} \mathrm{d}\mathbf{r} \int_{V_{\mathrm{j}}} \mathrm{d}\mathbf{r}' \frac{1}{|\mathbf{r} + \mathbf{R}_{\mathrm{i}} - \mathbf{R}_{\mathrm{j}} - \mathbf{r}'|} \left[ H_{\mathrm{i}\alpha;\mathrm{j}\alpha'}(\mathbf{r}, \mathbf{r}') \right] \Bigg\},
\end{aligned} \tag{3.55}$$

where

$$H_{\mathrm{i}\alpha;\mathrm{j}\alpha'}(\mathbf{r}, \mathbf{r}') = \rho_{\mathrm{i}\alpha}(\mathbf{r}) \rho_{\mathrm{j}\beta}(\mathbf{r}') - 2 Z_\alpha \delta(\mathbf{r}) \rho_{\mathrm{j}\beta}(\mathbf{r}') + Z_\alpha \delta(\mathbf{r}) Z_\beta \delta(\mathbf{r}') \tag{3.56}$$

At fixed $c_{\mathrm{i}\alpha}$, this expression satisfies the variational principles

$$\frac{\delta \Omega_c\, [\{c_{\mathrm{i}\alpha}\}]}{\delta \rho_{\mathrm{i}\alpha}(\mathbf{r})} = 0, \qquad \frac{\delta \Omega_c\, [\{c_{\mathrm{i}\alpha}\}]}{\delta \mathbf{M}_{\mathrm{i}\alpha}(\mathbf{r})} = 0, \tag{3.57}$$

*i.e.* it minimises the total energy with respect to both charge and magnetisation densities for fixed $\{c_{\mathrm{i}\alpha}\}$. This is in the same spirit as the variational principle for finite temperature density functional theory as described by Mermin [23]. The above $\Omega_c$ provides the explicit description for $\langle E_{\mathrm{int}} \rangle_0 [\{c_{\mathrm{i}\alpha}\}]$ needed in the linear response theory of Chap. 2 as outlined at the beginning of this chapter.

## 3.7 The Disordered Local Moment (DLM) Picture: Describing the Paramagnetic State

The CPA as described above should be viewed in the context of describing the electronic structure (and consequent internal energy) of a chemically disordered alloy. However, this is not the only type of disorder which can be described within this framework.

When modelling the phase behaviour of an alloy containing *magnetic* elements, such as Mn, Fe, Co, and Ni, it is necessary to consider how the magnetic state of a material impacts its electronic structure. In turn, one can imagine that an alloy's magnetic state could impact its propensity to order. In the formalism we have outlined so far, we have the option of describing the system with using either a non-magnetic or spin-polarised description. For the homogeneous, chemically disordered medium about which we intend to expand the free energy of the system, all lattice sites are equivalent, and a spin-polarised calculation in which magnetic moments emerge represents a description of the material's electronic structure at low temperatures. However, at finite temperature this picture of a magnetically ordered structure is erroneous and there will be significant thermal fluctuations. Indeed, at sufficiently high temperatures the alloy may become paramagnetic. A paramagnetic state is a state in which, although there are magnetic moments in the system, the overall magnetisation (on the average) is equal to zero. In light of the above discussion, it is necessary to find a satisfactory description of the material's electronic structure in this magnetically disordered regime.

The approach which will be employed in this work is the so-called disordered local moment (DLM) picture, which aims to describe the electronic and magnetic structure of itinerant electron systems at finite temperatures. Full details can be found in the seminal papers describing the theory [24–26], and in [27, 28]. Here we outline the key details necessary to understand how this approach will be used in our calculations. For brevity, in this outline, we will drop species label $\alpha$ and discuss the case of magnetic fluctuations for a single chemical species. (It is, of course, possible to include averaging over multiple chemical species in addition to averaging over magnetic disorder, if required.)

In a magnetic system, finite temperature and consequent thermal noise has two main effects. The first is that magnetic moments begin to rotate away from their ground state configuration. The second is that the occupancies of the electronic states are altered in accordance with the Fermi-Dirac distribution. Typically, the most important of these two effects is the first; the gradual disordering of the magnetic moment associated with each lattice site. It can be argued that magnetic moment fluctuations are slow on the timescale of electronic rearrangement, allowing a Born-Oppenheimer-like approximation to be made. We can therefore assume a given distribution of (normalised) magnetic moments $\{\hat{\mathbf{e}}_\mathrm{i}\}$ and solve the electronic structure problem for a given distribution. The magnetic moment $\hat{\mathbf{e}}_\mathrm{i}$ is given by integrating the normalised magnetisation density over the volume associated with the potential at site i,

$$\hat{\mathbf{e}}_{\mathrm{i}} = \frac{\int_{V_{\mathrm{i}}} \mathbf{M}(\mathbf{r})\mathrm{d}\mathbf{r}}{\left|\int_{V_{\mathrm{i}}} \mathbf{M}(\mathbf{r})\mathrm{d}\mathbf{r}\right|} \tag{3.58}$$

At $T = 0$, there are no thermal fluctuations, and the magnetic moments are in their ground state configuration. As the temperature increases, spin-wave-like excitations emerge as a consequence of thermal noise in the system, resulting in a spread of the distribution of magnetic moments, although there will still be a preferred direction along which they will point. Finally, above some critical temperature, the system will transition to a paramagnetic state, one in which the magnetisation, on the average, is equal to zero.

The DLM picture seeks to model the electronic structure of a magnetically disordered state by describing an optimal mean-field medium for a system of magnetic moments with distribution $P(\{\hat{\mathbf{e}}_{\mathrm{i}}\})$ which, in principle, can include detailed correlations between moments. At arbitrary temperature, this probability distribution is given by a Boltzmann distribution. The CPA condition for an impurity atom with moment $\hat{\mathbf{e}}_{\mathrm{i}_0}$ reads [28]

$$\prod_{\mathrm{i}} \int \mathrm{d}\Omega_{\mathrm{i}}\, \tau_{\mathrm{i}_0\mathrm{i}_0}(\{\hat{\mathbf{e}}_{\mathrm{i}}\}) P(\{\hat{\mathbf{e}}_{\mathrm{i}}\}) = \tau^{c}_{\mathrm{i}_0\mathrm{i}_0}, \tag{3.59}$$

*i.e.* we have replaced the sum weighted by concentrations of the chemically disordered case with an integral over solid angle weighted by the probability distribution of the moments. Evaluation of this for arbitrary temperature can be computationally demanding.

However, for the disordered alloy considered in Chap. 2 in which all lattice sites are equivalent, we can drop Latin index i. Further, above the Curie temperature of a material, in the paramagnetic state, where the probability distribution is uniform,

$$P(\hat{\mathbf{e}}) = \frac{1}{4\pi}, \tag{3.60}$$

the CPA condition is simplified to

$$\int \mathrm{d}\Omega\, \tau_{\mathrm{ii}}(\hat{\mathbf{e}}) = \tau^{c}_{\mathrm{ii}}. \tag{3.61}$$

By specifying a choice of global $\hat{\mathbf{z}}$ axis, taking advantage of the fact that the reference medium has zero overall spin polarisation, and following through some algebra transforming between local and global coordinate frames [24], it can be shown that this CPA condition is *entirely mathematically equivalent* to the CPA condition for an Ising alloy of 50% up spins and 50% down spins,[7] *i.e.*

[7] It should be emphasised that this equivalence *only* holds for the case of zero overall magnetisation in a system without spin-orbit coupling.

$$\frac{1}{2}\tau_{\mathrm{ii}}^{\uparrow} + \frac{1}{2}\tau_{\mathrm{ii}}^{\downarrow} = \tau_{\mathrm{ii}}^{c}, \tag{3.62}$$

where $\uparrow$ and $\downarrow$ denote a magnetic moment parallel or antiparallel to the global $\hat{\mathbf{z}}$ axis, respectively. This mathematical equivalence means that, within the CPA, we can model a truly paramagnetic state for roughly the same computational cost as for a conventional spin-polarised CPA calculation on a binary alloy.

In practice, one initialises a calculation of the 50:50 Ising system with magnetic moments and iterates to self-consistency to see which chemical species 'support' a local moment. Those which do not will converge to a state where both of the spin-up and spin-down Ising potentials have zero magnetisation. Local moments are sensitive to the crystal structure, lattice spacing, and their chemical environment. Fe can be relied upon to form a local moment in most cases, as can Mn. The local moments of Ni and Co are more elusive, while Cr tends not to form a local moment at all.

As an example of how the electronic structure of the paramagnetic state differs from that of a ferromagnetic state, in Sect. 3.9 we visualise the bandstructure and density of states for bcc Fe for both states and compare them. We shall see in Chaps. 7 and 8 how the magnetic state of alloy can alter its ordering tendencies.

## 3.8 The $S^{(2)}$ Theory for Multicomponent Alloys

From the preceding discussion, we now have a mechanism by which to describe the electronic structure of an alloy, using a non-magnetic, spin-polarised, or DLM description, as appropriate, from which we can obtain the self-consistent single-electron potentials and charge densities of DFT. We have already seen how the outlined approach can give us access to $E_{\mathrm{int}}[\{\xi_{\mathrm{i}\alpha}\}]$ and $\langle E_{\mathrm{int}}\rangle_0[\{c_{\mathrm{i}\alpha}\}]$. Finally, we can consider evaluation of the atomic short-range order,

$$S^{(2)}_{\mathrm{i}\alpha\mathrm{j}\alpha'} \equiv \left.\frac{\partial^2 \langle E_{\mathrm{int}}\rangle_0}{\partial c_{\mathrm{i}\alpha}\partial c_{\mathrm{j}\alpha'}}\right|_{\{c_\alpha\}}. \tag{3.63}$$

Recalling the Landau series expansion of the alloy free energy (Chap. 2, (2.16)), we now consider this second derivative in the context of density functional theory, where a fluctuation in concentrations results in an associated fluctuation of the electron density. We write $\bar{\Omega}$ to denote the average energy of the CPA medium, as we will be considering the response of this medium about the disordered reference state. For a set of small concentration fluctuations $\{\Delta c_{i\alpha}\}$, where the overall concentrations are unchanged,

$$\bar{\Omega}[\{c_{\mathrm{i}\alpha}\}] = \bar{\Omega}_0[\{c_\alpha\}] + \sum_{\mathrm{i}\alpha} \frac{\partial \bar{\Omega}}{\partial c_{\mathrm{i}\alpha}} \Delta c_{\mathrm{i}\alpha} + \sum_{\mathrm{i}\alpha} \sum_{\mathrm{k}\gamma} \int \mathrm{d}\mathbf{r}_{\mathrm{k}} \frac{\delta \bar{\Omega}}{\delta \rho_{\mathrm{k}\gamma}(\mathbf{r}_{\mathrm{k}})} \frac{\partial \rho_{\mathrm{k}\gamma}(\mathbf{r}_{\mathrm{k}})}{\partial c_{\mathrm{i}\alpha}} \Delta c_{\mathrm{i}\alpha}$$
$$+ \frac{1}{2} \sum_{\mathrm{i}\alpha} \sum_{\mathrm{j}\alpha'} \left[ \frac{\partial^2 \bar{\Omega}}{\partial c_{\mathrm{i}\alpha} \partial c_{\mathrm{j}\alpha'}} + \sum_{\mathrm{k}\gamma} \int \mathrm{d}\mathbf{r}_{\mathrm{k}} \frac{\partial}{\partial c_{\mathrm{k}\gamma}} \left( \frac{\delta \bar{\Omega}}{\delta \rho_{\mathrm{k}\gamma}(\mathbf{r}_{\mathrm{k}})} \right) \frac{\partial \rho_{\mathrm{k}\gamma}(\mathbf{r}_{\mathrm{k}})}{\partial c_{\mathrm{j}\alpha'}} \right] \Delta c_{\mathrm{i}\alpha} \, \Delta c_{\mathrm{j}\alpha'}. \tag{3.64}$$

As discussed in Chap. 2, the second and third terms of (3.64) vanish owing to fixed overall concentrations, $\sum_{\mathrm{i}} \Delta c_{\mathrm{i}\alpha} = 0$, and the DFT charge stationarity conditions, (3.57) respectively. The terms in square brackets (4th and 5th terms) constitute the direct correlation function, $S^{(2)}_{\mathrm{i}\alpha,\mathrm{j}\alpha'}$ [3]. It includes effects not only from how the single particle energies contribution (the first term of (3.55)) to the energy $\bar{\Omega}[\{c_{\mathrm{i}\alpha}\}]$ depends on the configurational configuration but also on how the electronic charge rearranges with fluctuations in composition. This latter piece depends on the charge responses $\frac{\partial \rho_{\mathrm{k}\gamma}(\mathbf{r}_{\mathrm{k}})}{\partial c_{\mathrm{j}\alpha'}}$.

The derivation of an expression for the term in square brackets, particularly in the multicomponent setting, is highly non-trivial. However, following the spirit of earlier work studying binary alloys [29, 30], this problem has been considered in detail by Khan, Staunton, and Stocks [3]. These authors derive a set of coupled equations for the variation of various key quantities of the theory which must be solved self-consistently and stitched together to obtain the ASRO. We consider these equations now.

### 3.8.1 Coupled Equations for $S^{(2)}_{\alpha\alpha'}(\mathbf{k})$

It is necessary to consider the variations of (3.42), (3.50), (3.46), and (3.55). To simplify the derivations, only systems with a monatomic basis are considered (*i.e.* i is replaced with $i$), while the response of the charge density, $\delta\rho_{i\alpha}(r)$, is expanded in terms of an orthonormal basis of Legendre polynomials, $f_n(r)$. These satisfy $\int f_n(r)dr = \delta_{n1}$ and $f_1(r) = 1$. This reduces the number of degrees of freedom from the number of points along a real-space grid to the number of basis functions considered. Any superscripts $n, m$ will refer to indices in this basis of Legendre polynomials.

The 'charge response' is defined via

$$\Phi^n_{i\alpha;j\beta} = \frac{\delta}{\delta v_{j\beta}} \int dr \rho_{i\alpha}(r) f_n(r) \mathrm{d}r, \tag{3.65}$$

which is a rectangular $s \times (s-1)$ matrix for an $s$ component alloy. Another approximation made is that $|\mathbf{r} + \mathbf{R}_i - \mathbf{R}_j - \mathbf{r}'| \to |\mathbf{R}_i - \mathbf{R}_j|$ when appropriate, which is equivalent to keeping only leading monopole terms for Coulomb interactions between pairs of cells. It is then possible to work with the 'site charges'

$$Q_{i\alpha} := \int_{V_0} \mathrm{d}\mathbf{r}[\rho_{i\alpha}(\mathbf{r}) - Z_\alpha \delta(\mathbf{r})], \tag{3.66}$$

'site polarisations'

$$P_i := \sum_\alpha \bar{c}_{i\alpha} Q_{i\alpha}, \tag{3.67}$$

and the lattice Fourier transform of the electrostatic interaction,

$$M(\mathbf{k}) := \sum_{i \neq 0} \frac{e^2}{|\mathbf{R}_i|} e^{-\mathbf{k}\cdot\mathbf{R}_i}. \tag{3.68}$$

Having made these approximations, before stating the set of coupled equations which must be solved to obtain $S^{(2)}_{\alpha\alpha'}$, it is first necessary to define relevant on-site quantities. These are given in terms of the Fermi-Dirac function $f(E-\mu)$, site impurity Green's function $G^\alpha$, site regular solution $Z_{\alpha;L}(\mathbf{r})$, and basis functions $f_n(r)$:

$$A^{mn}_\alpha := -\frac{1}{\pi} \int \mathrm{d}E\, f(E-\mu)\, \mathrm{Im} \int_{V_0} \mathrm{d}\mathbf{r} \int_{V_0} \mathrm{d}\mathbf{r}' f_m(\mathbf{r}) G^\alpha(E; \mathbf{r}, \mathbf{r}') \times \left[\frac{\mathrm{d}V_{\mathrm{xc}}}{\mathrm{d}\rho}(\rho_\alpha(\mathbf{r}')) f_n(\mathbf{r}') + \left(\int_{V_0} \mathrm{d}\mathbf{r}'' \frac{e^2}{|\mathbf{r}'-\mathbf{r}''|} f_n(\mathbf{r}'')\right)\right] G^\alpha(E; \mathbf{r}', \mathbf{r}) \tag{3.69}$$

$$B^{m}_\alpha := -\frac{1}{\pi} \int \mathrm{d}E\, f(E-\mu)\, \mathrm{Im} \int_{V_0} \mathrm{d}\mathbf{r} \int_{V_0} \mathrm{d}\mathbf{r}' f_m(\mathbf{r}) G^\alpha(E; \mathbf{r}, \mathbf{r}') G^\alpha(E; \mathbf{r}', \mathbf{r}) \tag{3.70}$$

$$F^n_{\alpha;LL'} := -\int_{V_0} \mathrm{d}\mathbf{r} Z_{\alpha;L}(\mathbf{r}) Z_{\alpha;L'}(\mathbf{r}) f_n(\mathbf{r}) \tag{3.71}$$

$$U^n_{\alpha;LL'} := -\int_{V_0} \mathrm{d}\mathbf{r} Z_{\alpha;L}(\mathbf{r}) Z_{\alpha;L'}(\mathbf{r}) \times \left[\frac{\mathrm{d}V_{\mathrm{xc}}}{\mathrm{d}\rho}(\rho_\alpha(\mathbf{r})) f_n(\mathbf{r}) + \left(\int_{V_0} \mathrm{d}\mathbf{r}' \frac{e^2}{|\mathbf{r}-\mathbf{r}'|} f_n(\mathbf{r}')\right)\right]. \tag{3.72}$$

Note that site index $i$ is dropped due to the equivalence of all sites in disordered solid solution.

It is also convenient to work with enlarged supermatrices with row (column) indices given by composite index $(L_1, L_2)$ for independent angular momentum indices $L_1, L_2$. Using this convention, it is possible to define CPA related supermatrices

$$\bar{\mathbb{D}}_{\alpha;L_1L_2;L_3L_4} := \bar{D}_{\alpha;L_1L_3} D_{\alpha;L_4L_2} \tag{3.73}$$

$$\mathbb{D}_{\alpha;L_1L_2;L_3L_4} := D_{\alpha;L_1L_3} \bar{D}_{\alpha;L_4L_2} \tag{3.74}$$

$$\mathbb{X}_{L_1L_2;L_3L_4} := \sum_\alpha \bar{c}_\alpha X_{\alpha;L_1L_3} X_{\alpha;L_4L_2} \tag{3.75}$$

$$\mathbb{C}_{L_1L_2;L_3L_4}(\mathbf{k}) := \frac{1}{V_{\mathrm{BZ}}} \int \mathrm{d}\mathbf{q}\, \Delta\tau^c(\mathbf{q})_{L_1L_3} \Delta\tau^c(\mathbf{q}-\mathbf{k})_{L_4L_2} \tag{3.76}$$

where the matrix $\Delta\tau^c(\mathbf{q}) = \tau^c(\mathbf{q}) - \tau^c_{00}$ and $\tau^c(\mathbf{q})$ is the Fourier transform of the CPA scattering path operator, $\tau^c_{ij}$. These matrices are used to describe the response of $t_\mu$ and $t_c$ matrices. Note that computation of $\mathbb{C}(\mathbf{k})$ is comparatively computationally expensive as it requires a convolution integral of the SPO $\tau^c(\mathbf{k})$ over the Brillouin zone.

Having made all of the above definitions, we can (at last) state a set of coupled equations defining the atomic short-range order, $\bar{\Psi}_{\alpha\alpha'}(\mathbf{k})$, and, consequently, $S^{(2)}_{\alpha\alpha'}$. They are:

$$\frac{\delta P}{\delta \nu_\gamma}(\mathbf{k}) := \sum_\sigma (Q_\sigma - Q_n)\beta\bar{\Psi}_{\sigma\gamma}(\mathbf{k}) + \sum_\sigma \bar{c}_\sigma \Phi^1_{\sigma\gamma}(\mathbf{k}), \tag{3.77}$$

$$\frac{\delta t_\mu^{-1}}{\delta \nu_\gamma}(\mathbf{k}) = \sum_n U^n_\mu \Phi^n_{\mu\gamma}(\mathbf{k}) + F^1_\mu M(\mathbf{k}) \frac{\delta P}{\delta \nu_\gamma}(\mathbf{k}), \tag{3.78}$$

$$[\mathbb{I} - \mathbb{X}\mathbb{C}(\mathbf{k})] \frac{\delta t_c^{-1}}{\delta \nu_\gamma}(\mathbf{k}) = [\mu] \sum (X_\mu - X_n)\beta\bar{\Psi}_{\mu\gamma}(\mathbf{k}) + \sum_\mu \bar{c}_\mu \bar{\mathbb{D}}_\mu \frac{\delta t_\mu^{-1}}{\delta \nu_\gamma}(\mathbf{k}), \tag{3.79}$$

$$\begin{aligned}\Phi^m_{\mu\gamma}(\mathbf{k}) = &\sum_n A^{mn}_\mu \Phi^n_{\mu\gamma}(\mathbf{k}) + B^m_\mu M(\mathbf{k}) \frac{\delta P}{\delta \nu_\gamma}(\mathbf{k}) \\ &- \frac{1}{\pi} \int \mathrm{d}E,\, f(E-\mu)\, \mathrm{Im} \sum_{LL'} F^m_{\mu;LL'} (\mathbb{D}_\mu \mathbb{C}(\mathbf{k}) \frac{\delta t_c^{-1}}{\delta \nu_\gamma}(\mathbf{k}))_{LL'},\end{aligned} \tag{3.80}$$

and

$$\begin{aligned}\beta\bar{\Psi}_{\mu\gamma}(\mathbf{k}) = \beta C_{\mu\gamma} + \sum_\sigma \beta C_{\mu\sigma} \Bigg\{ &\frac{1}{\pi} \int \mathrm{d}E f(E-\mu) \mathrm{Im}\mathrm{Tr} \left[ (X_\sigma - X_n) \mathbb{C}(\mathbf{k}) \frac{\delta t_c^{-1}}{\delta \nu_\gamma}(\mathbf{k}) \right] \\ &- (Q_\sigma - Q_n) M(\mathbf{k}) \frac{\delta P}{\delta \nu_\gamma}(\mathbf{k}) \Bigg\}.\end{aligned} \tag{3.81}$$

These equations must be solved self-consistently in a numerical fashion, but we will not concern ourselves with the details here.

Once the solution is obtained from the various CPA-related quantities, this set of equations define a means by which to obtain $S^{(2)}_{\alpha\alpha'}(\mathbf{k})$, and, therefore, $S^{(2)}_{i\alpha j\alpha'}$, as specified in (3.3). We will now briefly discuss the inclusion of the Onsager correction, as well as the band-only reduction, which represents a major simplification of the equations.

### 3.8.2 Including the Onsager Correction

It is naturally possible to consider the Onsager correction of Chap. 2, Sect. 2.7 in the context of these equations. Although $S^{(2)}_{\alpha\alpha'}(\mathbf{k})$ is not explicitly present in the above equations, the same logic of a shift $\Lambda_{\alpha\alpha'}$ applies.

Equation 3.81 is replaced with

$$\begin{aligned}\beta\hat{\Psi}_{\mu\gamma}(\mathbf{k}) =&\beta C_{\mu\gamma} + \sum_{\sigma}\beta C_{\mu\sigma}\Bigg\{\frac{1}{\pi}\int \mathrm{d}E f(E-\mu)\mathrm{ImTr}\Bigg[(X_\sigma - X_n)\mathbb{C}(\mathbf{k})\frac{\delta t_c^{-1}}{\delta v_\gamma}(\mathbf{k})\Bigg]\\&- (Q_\sigma - Q_n)M(\mathbf{k})\frac{\delta P}{\delta v_\gamma}(\mathbf{k}) - \sum_{\delta}{}' \Lambda_{\sigma\delta}\hat{\Psi}_{\delta\gamma}(\mathbf{k})\Bigg\},\end{aligned} \tag{3.82}$$

where

$$\begin{aligned}\Lambda_{\sigma\delta} :=& \frac{1}{V_{\mathrm{BZ}}}\int \mathrm{d}\mathbf{k}\sum_{\gamma}{}'\Bigg\{\frac{1}{\pi}\int \mathrm{d}E f(E-\mu)\mathrm{ImTr}\Bigg[(X_\sigma - X_n)\mathbb{C}(\mathbf{k})\frac{\delta t_c^{-1}}{\delta v_\gamma}(\mathbf{k})\Bigg]\\&- (Q_\sigma - Q_n)M(\mathbf{k})\frac{\delta P}{\delta v_\gamma}(\mathbf{k})\Bigg\}C^{-1}_{\gamma\delta}.\end{aligned} \tag{3.83}$$

It is also important to consider the response of the on-site charge density, $\rho_{\mathrm{j}\gamma}(\mathbf{r}_\mathrm{j})$. This needs to obey $\frac{\partial \rho_{\mathrm{j}\gamma}(\mathbf{r}_j)}{\partial c_{\mathrm{j}\alpha'}} = 0$, which is a consequence of the requirement that $\rho_{\mathrm{j}\gamma}(\mathbf{r}_\mathrm{j}) = \rho_{\mathrm{j}\gamma}(\mathbf{r}_\mathrm{j}; \{c_{\mathrm{i}\alpha}\}, \mathrm{i\,excluding\,j})$. In a similar manner to the restoration of the atomic short-range order above, we can restore the on-site charge response sum rule by replacing (3.80) with

$$\begin{aligned}\hat{\Phi}^m_{\mu\gamma}(\mathbf{k}) =& \sum_n A^{mn}_{\mu}\hat{\Phi}^n_{\mu\gamma}(\mathbf{k}) + B^m_{\mu}M(\mathbf{k})\frac{\delta P}{\delta v_\gamma}(\mathbf{k})\\&- \frac{1}{\pi}\int \mathrm{d}E\, f(E-\mu)\,\mathrm{Im}\sum_{LL'} F^m_{\mu;LL'}(\mathbb{D}_\mu\mathbb{C}(\mathbf{k})\frac{\delta t_c^{-1}}{\delta v_\gamma}(\mathbf{k}))_{LL'}\\&- \sum_{\sigma}{}' \Lambda^m_{\mu\delta}\hat{\Psi}_{\delta\gamma}(\mathbf{k}),\end{aligned} \tag{3.84}$$

where

$$\Lambda^m_{\mu\delta} := \frac{1}{V_{\rm BZ}} \int {\rm d}\mathbf{k} \sum_{\gamma}{}' \left\{ B^m_\mu M(\mathbf{k}) \frac{\delta P}{\delta \nu_\gamma}(\mathbf{k}) - \frac{1}{\pi} \int {\rm d}E\, f(E-\mu)\, {\rm Im} \sum_{LL'} F^m_{\mu;LL'} (\mathbb{D}_\mu \mathbb{C}(\mathbf{k}) \frac{\delta t_c^{-1}}{\delta \nu_\gamma}(\mathbf{k}))_{LL'} \right\} C^{-1}_{\gamma\delta}. \quad (3.85)$$

The combined restoration of the on-site sum rule for atomic short-range order and the on-site charge density will turn out to be important in what follows. In particular, in alloys where there are large transfers of electronic charge between elements (such as alloys of 3*d* and 5*d* transition metals), the restoration of the response of the on-site charge density will have a substantial impact on our estimated transition temperatures. An example is considered in Chap. 4, Sect. 4.3.1.

### 3.8.3 Band-Only Reduction

Equations 3.77–3.81 are highly complex, and a major simplification is presented by the so-called 'band-only' approximation. This imposes the requirement that there is no charge transfer, and no charge response. This means $Q_\alpha = Q_\beta$ and $\Phi^n_{\alpha\beta}(\mathbf{k}) = 0$. This can be shown to reduce (3.77)–(3.81) to a single equation,

$$\beta\bar{\Psi}_{\mu\gamma}(\mathbf{k}) = \beta C_{\mu\gamma} + \sum_{\sigma\nu}{}' \beta C_{\mu\sigma} \left\{ \frac{1}{\pi} \int {\rm d}E f(E-\mu) \times {\rm ImTr} \left[ (X_\sigma - X_n)\mathbb{C}(\mathbf{k}) \left[\mathbb{I} - \mathbb{X}\mathbb{C}(\mathbf{k})\right]^{-1} (X_\nu - X_n) \right] \right\} \beta\bar{\Psi}_{\nu\gamma}(\mathbf{k}). \quad (3.86)$$

By analogy with (2.55), the term in braces is identified as $S^{(2)}_{\sigma\nu}(\mathbf{k})$. Further, by analogy with (2.67), we obtain

$$\tilde{S}^{(2)}_{\alpha\alpha'}(\mathbf{k}) = \frac{1}{\pi} \int {\rm d}E f(E-\mu) {\rm ImTr} \left[ X_\alpha \mathbb{C}(\mathbf{k}) \left[\mathbb{I} - \mathbb{X}\mathbb{C}(\mathbf{k})\right]^{-1} X'_\alpha \right]. \quad (3.87)$$

Although the charge is frozen, the response to the band-term in the total energy is included (hence 'band-only'), and therefore band-related mechanisms such as van Hove singularities [31] and Fermi-surface nesting [32] are included. While this approximation is satisfactory for some alloys,[8] we will see later that for many of the alloys considered in this work the band-only approximation fails to obtain physical results. Some examples of this failure are given in Sects. 4.3.2 and 6.3.1. The inclusion of the effects of charge transfer and charge response due to an applied chemical

[8] *E.g.* for systems where constituent elements are isoelectronic and charge transfer between elements is small.

perturbation is similar to the consideration given in density functional perturbation theory (DFPT) [33, 34], used to describe lattice dynamics *ab initio* and response functions for phonons, *etc.*

## 3.9 Electronic Structure of Metals and Alloys: Illustrative Examples

Before concluding this chapter, let us consider the electronic structure of some prototypical metals and alloys, as modelled within the tennets of DFT. These examples will aid understanding and interpretation of results and arguments given in later chapters. All calculations are performed within the KKR formulation of DFT using the all-electron *Hutsepot* code [35]. When disordered systems are considered, the CPA is used to average over the disorder. Default settings are used for the self-consistent portion of the calculations, including a $20 \times 20 \times 20$ **k**-point mesh for Brillouin zone integrations and a 24-point semicircular Gauss-Legendre grid for contour integrals in the complex plane. Throughout, the local-density approximation is used, and the exchange-correlation functional is that of Perdew-Wang. Calculations are scalar-relativistic and, where relevant, spin-polarised. For these illustrative examples, we do not consider the effects of spin-orbit coupling, which would naturally be included in fully relativistic calculations, and would make some small differences to some of these calculations.

Before we begin, it should be emphasised that the electronic structure of metals and alloys is a vast subject area, and there is no possibility of covering it in depth in this work. For an introduction to solid state physics, the reader could consult the well-known books[9] of Kittel [36], Ashcroft and Mermin [37], or Simon [38]. For in depth details of band-theory, including applications to metals, a good reference would be the book of Singleton [39]. Finally, for specifics of metallic bonding and the structure of metals and alloys, a far better exposition than this author could possibly give is given by Pettifor [40].

### 3.9.1 *Simple Metals*

A so-called 'simple' metal is one whose conduction electrons come from *sp* shells. They are well-described by the Sommerfeld, or 'free electron' model, in which electron-electron and electron-ion interactions are ignored, and electrons are assumed to be 'free', with the only restriction placed upon them being that quantum state of the system can only be occupied by one electron, a consequence of the Pauli exclusion

[9] The author would like to express a preference for the book of Kittel; his own edition includes a chapter on alloy physics and spends far more time open on his desk than gathering dust on a bookshelf!

principle. The dispersion relation in three dimensions is then

$$E = \frac{\hbar^2 \mathbf{k}^2}{2m}, \tag{3.88}$$

and the density of states (DoS) is

$$g(E) \propto E^{\frac{1}{2}}. \tag{3.89}$$

An extension to this model is then the 'nearly-free electron model', in which there is a periodic but *very weak* potential, which turns out to be an excellent descriptor of alkali metals like Na and K [39]. Note that there is no description of spin in this model; we implicitly assume that each state can be filled by a spin-up electron and by a spin-down electron simultaneously. (The wavevector **k** and electron spin are both quantum numbers so the states ($\mathbf{k}$, $\uparrow$) and ($\mathbf{k}$, $\downarrow$) are different quantum states.)

As examples of simple metals, we consider Al, forming on the fcc lattice, and Na, forming on the bcc lattice. Visualised in Fig. 3.2 are the bandstructure and associated DoS of Na and Al. The reduced zone scheme is used[10]: every wavevector **k** is mapped back into the first Brillouin zone by translating with a reciprocal lattice vector **G**. The bandstructures are computed using the *Hutsepot* code with cubic lattice parameters of 4.225 and 4.05 Å respectively, consistent with their experimental values [36]. The parabolic nature of the bands can be clearly seen in both figures, and it can also be seen that $g(E) \sim \sqrt{E}$, as expected.

### 3.9.2 Non-magnetic Transition Metals

Transition metals are generally classified to be those metals lying in the $d$-block of the periodic table, who have $d$-like valence electrons. Electrons in $d$-orbitals are, typically, more localised than those in $s$ and $p$ orbitals, and they contribute narrow $d$-bands to a material's bandstructure. The bandwidth is broadened as we move down the table and the principal quantum number, $n$, is increased; the 3$d$ bands are typically far narrower than 5$d$ bands.

Visualised in Fig. 3.3 are the bandstructures of Cu (a 3$d$ transition metal) and Au (a 5$d$ transition metal). Again, these are computed using the *Hutsepot* code, with lattice parameters of 3.61 and 4.08 Å respectively, consistent with their experimental values [36]. Strictly, the scalar-relativistic calculation performed here is not quite sufficient for Au, which is heavy enough for spin-orbit effects to become significant [41] and results in a need for a calculation which is either fully relativistic or in which spin-orbit effects are included in some perturbative manner [42–45]. However,

[10] The three schemes for plotting functions in reciprocal space are the extended zone scheme, the reduced zone scheme, and the periodic zone scheme, and an overview is given in many textbooks, *e.g.* [36] or [38].

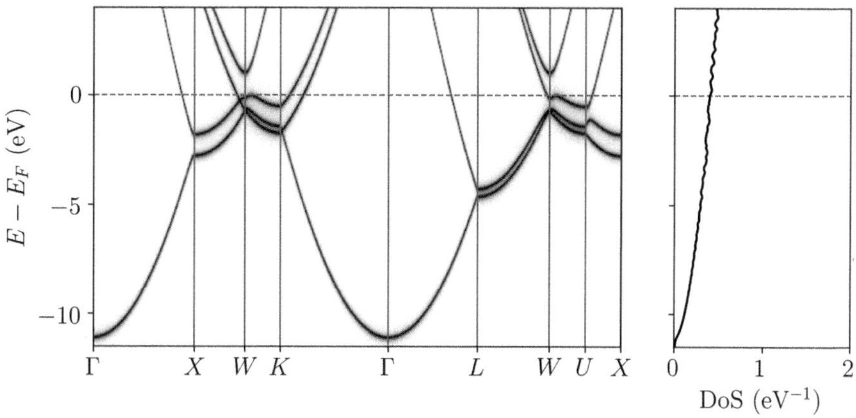


(a) Al

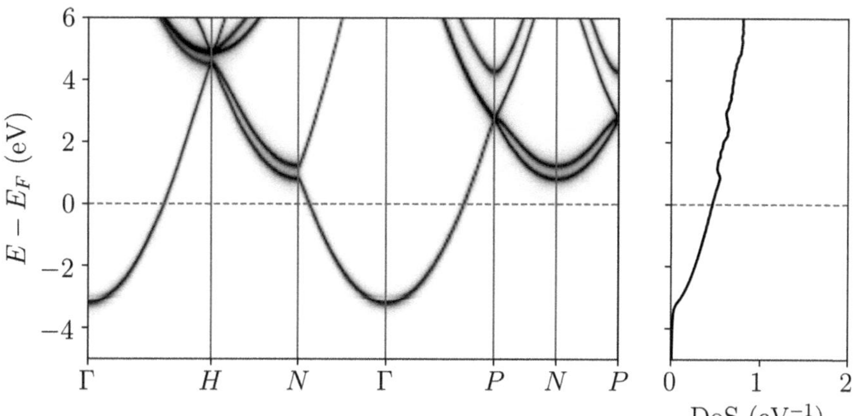


(b) Na

**Fig. 3.2** Visualisation of the bandstructure and associated DoS for elemental Al on the fcc lattice, and for elemental Na on the bcc lattice. Note that the slight blurring of the bandstructure is an artefact of the small imaginary component of the energy used in KKR calculations of a Bloch spectral function; the bands are truly delta functions

this elementary calculation captures enough of the relevant features of the electronic structure for our purposes.

Examining the bandstructures and associated density of states for fcc Cu and Au visualised in Fig. 3.3, we see that at both low and high energies (towards the bottom and top of both plots), the bands are broad in energy and close to parabolic in character, similar to the bands of Al. These near-parabolic bands are associated with $sp$-like electrons and make a notable contribution to the total DoS across a broad energy range. At intermediate energies, a collection of much narrower bands emerge; these are associated with the $d$ electrons of these systems. Electrons in $d$-orbitals are more localised than those in $sp$ shells, which reduces the overlap

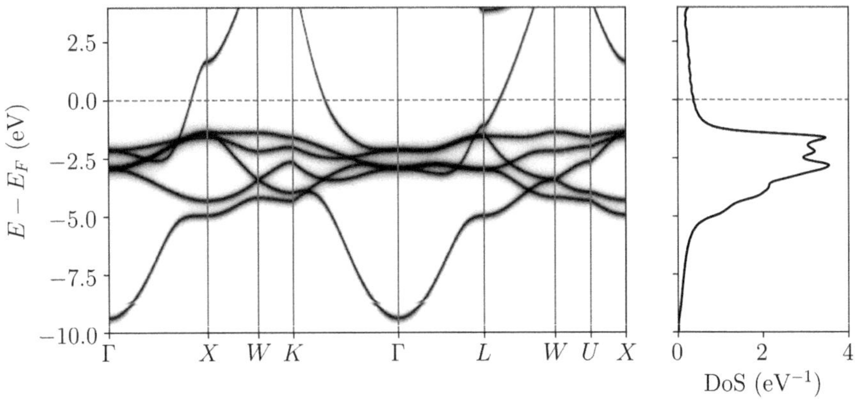


(a) Cu

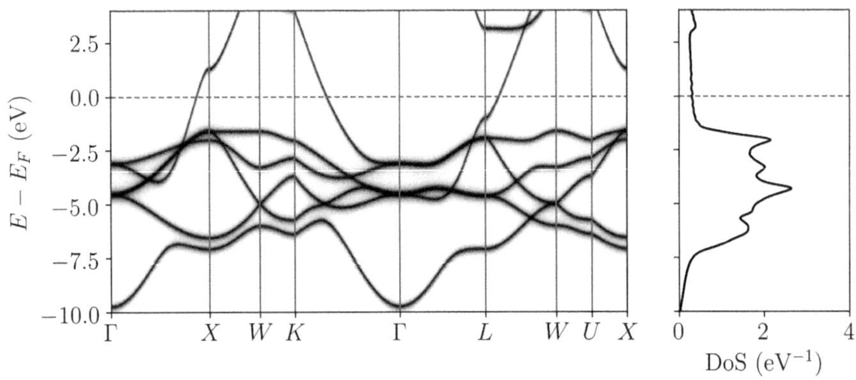


(b) Au

**Fig. 3.3** Visualisation of the bandstructure and associated DoS for elemental Cu and Au on the fcc lattice. The difference in width between the 3*d* bands of Cu and the 5*d* bands of Au can clearly be seen

between states on adjacent atoms. In turn, this flattens the bands associated with these *d* electrons, as electrons are able to occupy states very close in energy to one another. That said, there is a clear visible difference between the narrow 3*d* bands associated with Cu and the broader 5*d* bands of Au. Again, this is understood in terms of localisation of states; electrons in 3*d* orbitals are 'closer' to the nucleus and have reduced screening from the core electrons, resulting in highly localised states and narrow bands. By contrast, electrons in 5*d* orbitals are 'further' from the nucleus and experience increased screening, resulting in electronic states less localised in real space, resulting in broader bands. These differences in bandwidth will turn out to be important in discussing the origins of ordering tendencies in systems where elements from different rows of the *d* block are alloyed, such as in Chap. 4.

### 3.9.3 Magnetic Transition Metals

The previous examples have considered elements in which a spin-polarised treatment of the electronic structure is not necessary. We now consider Fe on a bcc lattice, for which spin-polarisation of the electronic structure is essential, as Fe is one of the prototypical ferromagnetic systems [46]. In Fig. 3.4a, we visualise the bandstructure and DoS for bcc Fe modelled evaluated via a ferromagnetic, spin-polarised calculation resolved into its spin-up and spin-down channels. The net magnetic moment on Fe is apparent when the number of states below $E_F$ is considered in both spin

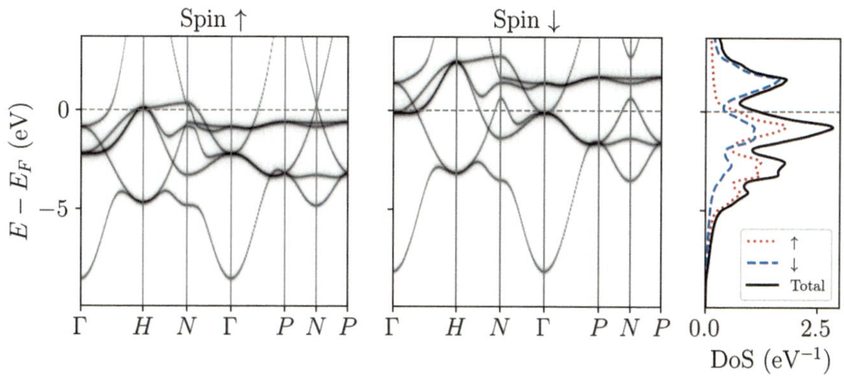


(a) Spin-polarised bandstructure and DoS for ferromagnetic Fe on the bcc lattice.

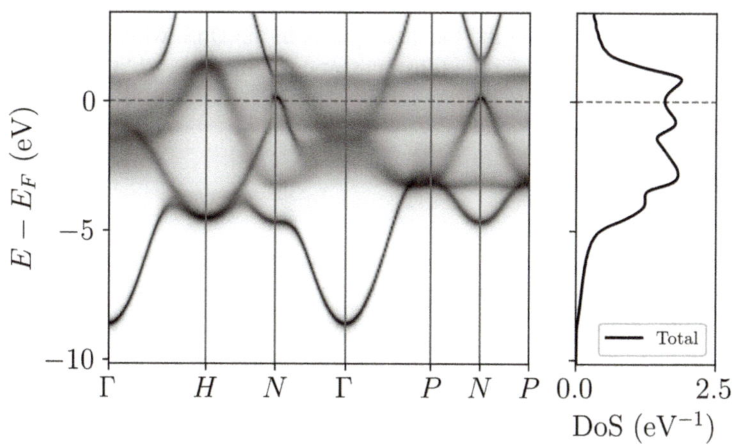


(b) Spin-polarised bandstructure and DoS for paramagnetic Fe on the bcc lattice, modelled within the DLM picture.

**Fig. 3.4** Visualisation of the Bloch spectral function and associated DoS for a CuAu alloy in both disordered (A1) and ordered ($L1_0$) phases, both with an underlying fcc lattice

channels. There are more spin-up states than spin-down, resulting in a net magnetic moment. It can also be seen that the overall profile of the DoS curves, as well as the bands themselves in both channels have a similar shape but are shifted in energy relative to one another, indicative of a mechanism similar to a rigid band shift at play. When the sum of both spin channels is considered, forming the total DoS for the ferromagnetic calculation, we see evidence of spin-polarisation remains in the form of some remaining finer detail in the DoS.

We then perform a paramagnetic calculation for bcc Fe modelled in its paramagnetic state as described within the DLM picture [24–26] as discussed in Sect. 3.7. This calculation is *still* spin-polarised, but the overall magnetic moment is zero, consistent with the picture of the paramagnetic state of a metal. It is only meaningful to consider the total DoS in this case, as there is a net-zero magnetic moment on the average. In addition, due to the on-site disorder and broken translational symmetry in the paramagnetic state, it is no longer meaningful to talk about a true 'bandstructure'. However, because there is still an underlying crystal lattice, something close to a bandstructure can be recovered in the form of the Bloch spectral function, a key quantity in the KKR-CPA formulation of electronic structure. The substitutional disorder broadens the Dirac delta functions of the original bandstructure into Lorentzians with finite width. Note that not all bands are smeared by an equal amount; the disorder has a more pronounced effect on some electronic states than others. In the case of paramagnetic Fe, it is the partially filled $d$-bands associated with the system's magnetic moments that are most heavily smeared.

The BSF and total DoS are visualised in Fig. 3.4b. As discussed, compared to the ferromagnetic calculation, sharp features in the DoS have been smeared out, a direct consequence of the CPA averaging over the probability distribution for the magnetic moments. However, a local spin-polarisation of the electronic structure is still evident, as can be seen in the localised peaks in the DoS, which are at similar energies as for the total DoS for the ferromagnetic calculation.

### 3.9.4 Binary Alloys

In a similar manner to a magnetically disordered system, in a disordered alloy (a solid solution) translational symmetry is broken due to the random substitution of elements on the lattice sites, and it is again no longer meaningful to talk about a true 'bandstructure'. Again, though, the BSF recovers something similar to a bandstructure. The finite width of the Lorentzians, coming from the smeared out bands, can be related to the finite lifetime of electronic states in a disordered alloy. (A disordered alloy can still has a finite 'residual' resistivity at low temperatures due to these finite lifetimes.) The broader the smearing of a band, the shorter the associated electronic lifetime.

As an example, we consider disordered, fcc CuAu, for which the Bloch spectral function and DoS are visualised in Fig. 3.5a. The calculation is scalar-relativistic, and the lattice parameter used is the weighted average of the two lattice parameters of

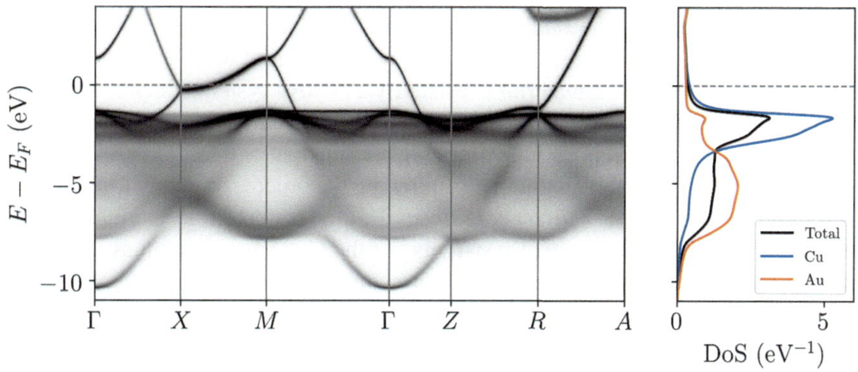


(a) Disordered (A1) CuAu: smeared out bandstructure.

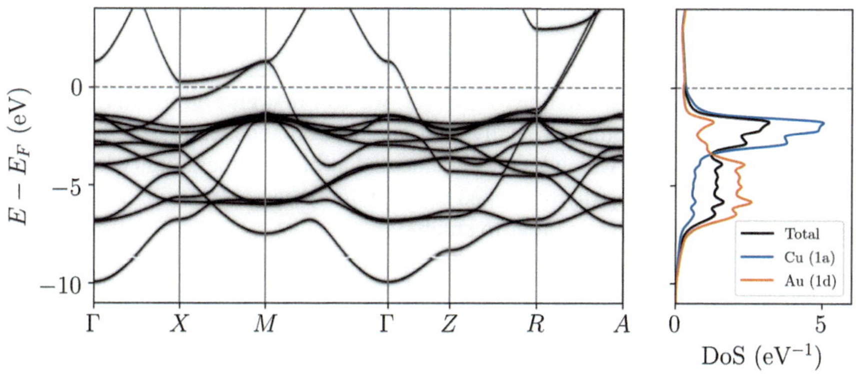


(b) Perfect $L1_0$ ordering: full bandstructure.

**Fig. 3.5** Visualisation of the Bloch spectral function and associated DoS for a CuAu alloy in both disordered (A1) and ordered ($L1_0$) phases, both with an underlying fcc lattice

Cu and Au. Here, we use the simple tetragonal representation of the fcc lattice, with a two-atom basis, for consistency with a calculation to follow. The Bloch spectral function is therefore visualised around the IBZ of the simple tetragonal lattice. The smearing out of the bandstructure can clearly be seen, as can hybridised bonding states forming at low energies, indicated by the slight reduction in energy of the lower-lying bands.

When an alloy orders and forms an intermetallic phase,[11] translational symmetry is restored, and a true bandstructure is recovered. Visualised in Fig. 3.5b is the bandstructure of $L1_0$ CuAu an ordered structure on the fcc lattice consisting of alternate layers of Cu and Au atoms. This structure is visualised in Fig. 2.2. A bandstructure is recovered, and the associated density of states plot has much finer detail than the disordered structure.

[11] An intermetallic phase is a regular, periodic arrangement of metal atoms.

A direct example of the implications of these calculations are given by experimental measurements of the residual resistivity of a material, which is defined as the low-temperature limit of a material's resistivity. A CuAu system which is quenched from high temperature, freezing in the disordered arrangement of atoms, has a residual resistivity which is almost three times higher than that of the same system which has been annealed and cooled gradually, allowing the ordered structure to form [36]. This reduced resistivity is a *direct consequence* of the restored translational symmetry. Visualisations of the residual resistivity for the $Cu_xAu_{1-x}$ system as a function of $x$ for both quenched and annealed samples are given in [36, 47].

### 3.9.5 Magnetic Alloys

It is possible to simulate both magnetic and compositional disorder by using the CPA to average over both compositional and magnetic disorder simultaneously, as will be illustrated in our final example. However, we will first give an intuitive example of a system which is compositionally *disordered* but magnetically *ordered*. This is reflective of a material which has been annealed at high temperature (where atoms are disordered), then rapidly quenched to a low temperature, freezing the atoms in place but allowing magnetic order to emerge.

The example considered here is FeNi, a system known to be particularly magnetically soft in its A1 (atomically disordered) phase, although surprisingly magnetically hard in its atomically ordered $L1_0$ phase [48–50]. The BSF and spin- and species-resolved DoS for both the A1 and $L1_0$ phases are visualised in Fig. 3.6. Again, we work in the two-atom representation of the $L1_0$ ordered phase, so the IBZ is that of a simple tetragonal lattice.

Considering first the atomically disordered (A1) phase, Fig. 3.6a, we see how the compositional disorder has a much greater impact on states in the minority spin channel than in the majority, where the $d$-bands are almost completely filled. As a consequence, we might expect the majority of electrical current in a system such as this to be carried by spin-up electrons; sharp bands correspond to a large electronic mean-free-path, while smeared-out bands correspond to an electronic mean-free-path which is short. The chemical ordering into the $L1_0$ phase, Fig. 3.6b removes this smearing and, as for the CuAu system, a full bandstructure is restored, albeit with the inevitable exchange splitting as a consequence of the magnetic ordering.

### 3.9.6 High-Entropy Alloys

Finally, we note that, in the setting of multicomponent alloys, the effects of disorder can be particularly profound. In Fig. 3.7, we visualise the BSF and associated DoS for the Cantor alloy, CrMnFeCoNi, modelled in its paramagnetic state via the DLM picture. (Atomic ordering tendencies in the Cantor alloy will be discussed in Chaps. 6

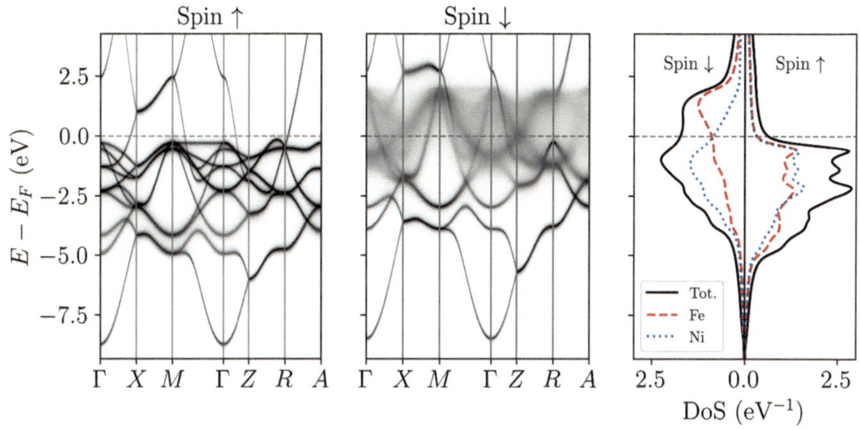


**(a) Atomically disordered (A1): smearing of bandstructure primarily in minority-spin channel.**

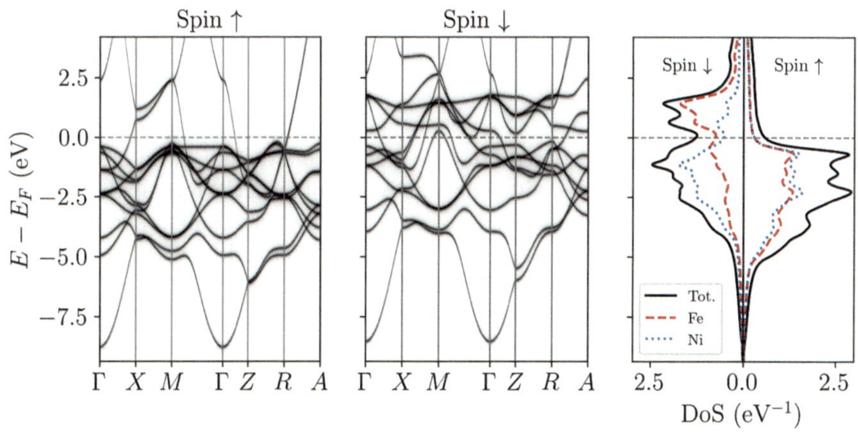


**(b) Perfect $L1_0$ atomic ordering: full bandstructure recovered.**

**Fig. 3.6** Visualisation of the Bloch spectral function and associated DoS for the ferromagnetic, equiatomic FeNi alloy in both atomically disordered (A1) and atomically ordered ($L1_0$) phases

and 7.) We see that, particularly around the Fermi level, $E_F$, there is huge smearing of electronic states.

Indeed, smearing such as the above can be observed experimentally. For example, [51] catalogues the effect of chemical disorder on the bandstructure and Fermi-surface of the equiatomic CrFeCoNi system, comparing KKR-CPA predictions with experimental observations. Both the bandstructure and Fermi surface are found be be highly smeared, with the Fermi surface being having a smearing as large as 40% of $2\pi/a$. Excellent agreement is found between theory and experiment, validating the effectiveness of the KKR-CPA approach when modelling these complex solid solutions.

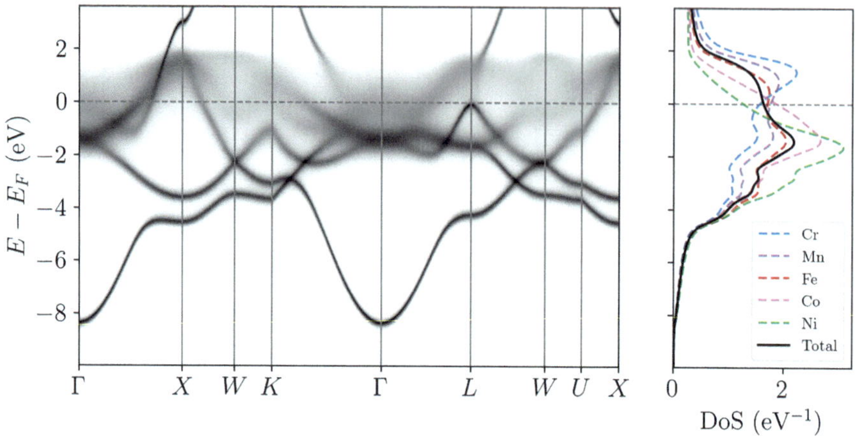


**Fig. 3.7** Visualisation of the Bloch spectral function and associated DoS for the CrMnFeCoNi alloy in its paramagnetic state, as modelled within the DLM picture

## 3.10 Summary

In this chapter, we have outlined how the electronic structure of an alloy can be described from first principles within the framework of the KKR formulation of DFT. This enables evaluation of the internal energy of the alloy for a configuration specified with the occupation numbers, $\{\xi_{i\alpha}\}$. We have then described the CPA, a means by which to evaluate a mean-field estimate of the internal energy of a disordered alloy with average site occupancies $\{c_{i\alpha}\}$. The DLM picture has also been outlined to describe how to model the paramagnetic state of an alloy, which is the magnetic state most relevant to the temperatures at which materials processing occurs. We then outlined the $S^{(2)}$ theory for multicomponent alloys as derived by Khan, Staunton, and Stocks [3], and highlighted how the inclusion of the effects of charge-transfer and charge-response set it apart from some alternative CPA-based approaches. Finally, we presented some illustrative examples of the electronic structure of some prototypical metals and alloys. It remains to describe the results of the implementation and analysis of the $S^{(2)}$ theory when applied to a number of example systems.

## References


1. P. Hohenberg, W. Kohn, Phys. Rev. **136**(3B), B864 (1964). https://doi.org/10.1103/PhysRev.136.B864
2. W. Kohn, L.J. Sham, Phys. Rev. **140**(4A), A1133 (1965). https://doi.org/10.1103/PhysRev.140.A1133
3. S.N. Khan, J.B. Staunton, G.M. Stocks, Phys. Rev. B **93**(5), 054206 (2016). https://doi.org/10.1103/PhysRevB.93.054206

4. R.M. Martin, *Electronic Structure: Basic Theory and Practical Methods* (Cambridge University Press, Cambridge, UK, 2004)
5. F. Giustino, *Materials Modelling Using Density Functional Theory: Properties and Predictions* (Oxford University Press, Oxford; New York, 2014)
6. J.P. Perdew, K. Schmidt, AIP Conf. Proc. **577**(1), 1 (2001). https://doi.org/10.1063/1.1390175
7. R.O. Jones, O. Gunnarsson, Rev. Modern Phys. **61**(3), 689 (1989). https://doi.org/10.1103/RevModPhys.61.689
8. R.G. Parr, W. Yang, *Density-Functional Theory of Atoms and Molecules*. International Series of Monographs on Chemistry (Oxford University Press, New York, NY, 1994)
9. D.M. Ceperley, B.J. Alder, Phys. Rev. Lett. **45**(7), 566 (1980). https://doi.org/10.1103/PhysRevLett.45.566
10. J.P. Perdew, Y. Wang, Phys. Rev. B **45**(23), 13244 (1992). https://doi.org/10.1103/PhysRevB.45.13244
11. P. Strange, *Relativistic Quantum Mechanics: With Applications in Condensed Matter and Atomic Physics*, 1st edn. (Cambridge University Press, 1998). https://doi.org/10.1017/CBO9780511622755
12. J. Kübler, *Theory of Itinerant Electron Magnetism*. International Series of Monographs on Physics (Oxford University Press, Oxford; New York, 2009)
13. A. Gonis, W.H. Butler, *Multiple Scattering in Solids*. Graduate Texts in Contemporary Physics (Springer, New York, 2000). https://www.springer.com/gb/book/9780387988535
14. J. Zabloudil (ed.), *Electron Scattering in Solid Matter: A Theoretical and Computational Treatise*. No. 147 in Springer series in solid-state sciences (Springer, Berlin; New York, 2005)
15. J.S. Faulkner, G.M. Stocks, Y. Wang, *Multiple Scattering Theory: Electronic Structure of Solids*, 1st edn. (IOP Publishing, Bristol, UK, 2018). https://doi.org/10.1088/2053-2563/aae7d8
16. A. Gonis, *Green Functions for Ordered and Disordered Systems*. Studies in Mathematical Physics (North-Holland, Amsterdam; New York, 1992)
17. P. Soven, Phys. Rev. **156**(3), 809 (1967). https://doi.org/10.1103/PhysRev.156.809
18. G.M. Stocks, W.M. Temmerman, B.L. Gyorffy, Phys. Rev. Lett. **41**(5), 339 (1978). https://doi.org/10.1103/PhysRevLett.41.339
19. J.S. Faulkner, G.M. Stocks, Phys. Rev. B **21**(8), 3222 (1980). https://doi.org/10.1103/PhysRevB.21.3222
20. A. Zunger, S.H. Wei, L.G. Ferreira, J.E. Bernard, Phys. Rev. Lett. **65**(3), 353 (1990). https://doi.org/10.1103/PhysRevLett.65.353
21. D.D. Johnson, D.M. Nicholson, F.J. Pinski, B.L. Györffy, G.M. Stocks, Phys. Rev. B **41**(14), 9701 (1990). https://doi.org/10.1103/PhysRevB.41.9701
22. R. Zeller, J. Phys.: Condens. Matter **16**(36), 6453 (2004). https://doi.org/10.1088/0953-8984/16/36/011
23. N.D. Mermin, Phys. Rev. **137**(5A), A1441 (1965). https://doi.org/10.1103/PhysRev.137.A1441
24. A.J. Pindor, J. Staunton, G.M. Stocks, H. Winter, J. Phys. F: Metal Phys. **13**(5), 979 (1983). https://doi.org/10.1088/0305-4608/13/5/012
25. J. Staunton, B. Gyorffy, A. Pindor, G. Stocks, H. Winter, J. Magn. Magn. Mater. **45**(1), 15 (1984). https://doi.org/10.1016/0304-8853(84)90367-6
26. B.L. Gyorffy, A.J. Pindor, J. Staunton, G.M. Stocks, H. Winter, J. Phys. F: Metal Phys. **15**(6), 1337 (1985). https://doi.org/10.1088/0305-4608/15/6/018
27. E. Mendive-Tapia, J.B. Staunton, Phys. Rev. Lett. **118**(19), 197202 (2017). https://doi.org/10.1103/PhysRevLett.118.197202
28. E. Mendive Tapia, *Ab initio Theory of Magnetic Ordering: Electronic Origin of Pair- and Multi-Spin Interactions*. Springer Theses (Springer, Cham, 2020). http://link.springer.com/10.1007/978-3-030-37238-5
29. J.B. Staunton, D.D. Johnson, F.J. Pinski, Phys. Rev. B **50**(3), 1450 (1994). https://doi.org/10.1103/PhysRevB.50.1450
30. D.D. Johnson, J.B. Staunton, F.J. Pinski, Phys. Rev. B **50**(3), 1473 (1994). https://doi.org/10.1103/PhysRevB.50.1473

31. J.F. Clark, F.J. Pinski, D.D. Johnson, P.A. Sterne, J.B. Staunton, B. Ginatempo, Phys. Rev. Lett. **74**(16), 3225 (1995). https://doi.org/10.1103/PhysRevLett.74.3225
32. E. Bruno, B. Ginatempo, E. Sandro Giuliano, Phys. Rev. B **63**(17), 174107 (2001). https://doi.org/10.1103/PhysRevB.63.174107
33. S.Y. Savrasov, Phys. Rev. B **54**(23), 16470 (1996). https://doi.org/10.1103/PhysRevB.54.16470
34. S. Baroni, S. de Gironcoli, A. Dal Corso, P. Giannozzi, Rev. Modern Phys. **73**(2), 515 (2001). https://doi.org/10.1103/RevModPhys.73.515
35. M. Hoffmann, A. Ernst, W. Hergert, V.N. Antonov, W.A. Adeagbo, R.M. Geilhufe, H. Ben Hamed, Phys. Status Solidi (b) **257**(7), 1900671 (2020). https://doi.org/10.1002/pssb.201900671
36. C. Kittel, *Introduction to Solid State Physics*, 8th edn. (Wiley, Hoboken, NJ, 2005)
37. N.W. Ashcroft, N.D. Mermin, *Solid State Physics* (Holt, Rinehart and Winston, New York, 1976)
38. S.H. Simon, *The Oxford Solid State Basics*, 1st edn. (Oxford University Press, Oxford, 2013)
39. J. Singleton, *Band Theory and Electronic Properties of Solids*. Oxford Master Series in Condensed Matter Physics (Oxford University Press, Oxford; New York, 2001)
40. D.G. Pettifor, *Bonding and Structure of Molecules and Solids* (Oxford University Press, Oxford; New York, 1995)
41. J.B. Staunton, Rep. Prog. Phys. **57**(12), 1289 (1994). https://doi.org/10.1088/0034-4885/57/12/002
42. J. Staunton, B.L. Gyorffy, P. Weinberger, J. Phys. F: Metal Phys. **10**(12), 2665 (1980). https://doi.org/10.1088/0305-4608/10/12/009
43. P. Weinberger, J. Staunton, B.L. Gyorffy, J. Phys. F: Metal Phys. **12**(10), 2229 (1982). https://doi.org/10.1088/0305-4608/12/10/017
44. P. Strange, J. Staunton, B.L. Gyorffy, J. Phys. C: Solid State Phys. **17**(19), 3355 (1984). https://doi.org/10.1088/0022-3719/17/19/011
45. P. Strange, H. Ebert, J.B. Staunton, B.L. Gyorffy, J. Phys.: Condens. Matter **1**(18), 2959 (1989). https://doi.org/10.1088/0953-8984/1/18/002
46. J. Staunton, B.L. Gyorffy, A.J. Pindor, G.M. Stocks, H. Winter, J. Phys. F: Metal Phys. **15**(6), 1387 (1985). https://doi.org/10.1088/0305-4608/15/6/019
47. C.H. Johansson, J.O. Linde, Annalen der Physik **417**(1), 1 (1936). https://doi.org/10.1002/andp.19364170102
48. L. Néel, J. Pauleve, R. Pauthenet, J. Laugier, D. Dautreppe, J. Appl. Phys. **35**(3), 873 (1964). https://doi.org/10.1063/1.1713516
49. L.H. Lewis, A. Mubarok, E. Poirier, N. Bordeaux, P. Manchanda, A. Kashyap, R. Skomski, J. Goldstein, F.E. Pinkerton, R.K. Mishra, R.C. Kubic Jr., K. Barmak, J. Phys.: Condens. Matter **26**(6), 064213 (2014). https://doi.org/10.1088/0953-8984/26/6/064213
50. C.D. Woodgate, C.E. Patrick, L.H. Lewis, J.B. Staunton, J. Appl. Phys. **134**(16), 163905 (2023). https://doi.org/10.1063/5.0169752
51. H.C. Robarts, T.E. Millichamp, D.A. Lagos, J. Laverock, D. Billington, J.A. Duffy, D. O'Neill, S.R. Giblin, J.W. Taylor, G. Kontrym-Sznajd, M. Samsel-Czeka la, H. Bei, S. Mu, G.D. Samolyuk, G.M. Stocks, S.B. Dugdale, Phys. Rev. Lett. **124**(4), 046402 (2020). https://doi.org/10.1103/PhysRevLett.124.046402

# Chapter 4
# Atomic Short-Range Order and Phase Stability of the Refractory High-Entropy Alloys

**Abstract** We apply our perturbative, DFT-based approach for analysing emergent atomic short- and long-range order in multicomponent alloys to the refractory high-entropy alloys (HEAs). This is a family of HEAs forming on the bcc lattice and containing the elements V, Nb, Mo, Ta, and W, which are of interest for applications where materials are subjected to high temperatures and levels of radiation. We choose to study the prototypical refractory HEAs, VNbMoTaW and NbMoTaW, as well as selected subsystems. Our modelling approach predicts a B2 ordering in the NbMoTaW system while, in the VNbMoTaW system, the addition of V drives a competing, B32 ordering. These ordering tendencies are interpreted physically as originating in electronic mechanisms including atomic size and bandwidth differences, charge transfer between elements, and valence differences between chemical species. Lattice-based Monte Carlo simulations using an effective pair interaction enable the nature of atomic ordering below the initial disorder-order transition to be studied in detail. The results are shown to compare favourably with alternative modelling approaches, and with experimental data. We also examine the failure of the band-only approximation to describe these systems, as well as discuss the importance of the Onsager correction.

## 4.1 Introduction

As discussed briefly in Chap. 1, a relatively recent development in the field of materials science is the discovery of 'multicomponent,' 'multiprinciple element,' or 'high-entropy' alloys (HEAs). Traditionally, work on alloys focussed on systems composed of one or two principal elements with relatively small additions of other elements to enhance desired properties. In 2004, however, seminal papers by Cantor *et al.* [1] and Yeh *et al.* [2] demonstrated that it was sometimes possible to combine multiple elements with similar atomic sizes in roughly equal concentrations to form stable, single-phase solid solutions. It was Yeh *et al.* who attributed the stability of the solid solution to a large contribution to the Gibbs free energy of the system from the configurational entropy, coining the term 'high-entropy alloy.' HEAs are known from both theoretical calculations and experiments to often have superior physical properties



C. D. Woodgate, *Modelling Atomic Arrangements in Multicomponent Alloys*,
Springer Series in Materials Science 346,
https://doi.org/10.1007/978-3-031-62021-8_4

compared to the base elements from which they are synthesised, and it often holds true that the whole is greater than the sum of its parts. Some improved properties of HEAs over pure elements or binary alloys include exceptional hardness, high tensile strength, and excellent resistance to wear, irradiation, and fracture [3–7].

One of the canonical families of HEAs are those based on the refractory elements: V, Nb, Mo, Ta, and W. First synthesised by Senkov *et al.* in 2010 [8, 9], this class of HEAs are of particular interest for applications in areas such as fission and fusion power, because the base elements from which they are produced have extremely high melting temperatures and are resistant to being irradiated [10–12].

The two original equiatomic refractory HEAs are NbMoTaW and VNbMoTaW, and we elect to study these along with the three component, medium-entropy, NbMoTa, all of which are known to form single-phase solid solutions on a bcc lattice [8, 9, 13]. A variety of approaches have been used to study compositional order in these materials and have highlighted interesting behaviours, suggesting they warrant further investigation [10, 14–22]. In addition, this existing literature will allow us to compare the results of our approach with results obtained using alternative methods.

This chapter is structured as follows. In Sect. 4.2 we will begin by examining the electronic structure of the atomically disordered, solid solutions, to pull out features that will be relevant to atomic ordering in these materials. Rather than just studying the multicomponent systems, we will also examine selected binary subsystems for comparison. Proceeding, in Sect. 4.3, results will be presented from the linear response theory outlined in Chaps. 2 and 3 applied to the aforementioned multicomponent and binary refractory alloys, with insight as to the physical origins of ordering elucidated in terms of the materials' electronic structure. In this section we will illustrate the importance of including the Onsager correction when modelling these systems, and also demonstrate the failure of the band-only approximation. Proceeding, in Sects. 4.4 and 4.5 we fit to a pairwise Hamiltonian and perform atomistic modelling to further explore the phase space of these systems. In Sect. 4.7, the results are summarised and an outlook given.

## 4.2 Electronic Structure

As a starting point for linear response calculations and analysis of the phase behaviour of these systems, we first perform self-consistent DFT calculations, as outlined in Chap. 3, to obtain the single-electron potentials for the disordered alloys. The potentials are generated in the KKR formulation of DFT, using the CPA to produce an effective medium reflecting the average electronic structure of the high-temperature, high-symmetry, disordered solid solution [23–25]. We use the all-electron *Hutsepot* code [26] to generate these potentials although, in principle, any KKR-CPA code would also be suitable. We perform scalar-relativistic calculations within the ASA [27] with an angular momentum cutoff of $l_{\max} = 3$ for basis set expansions, a $20 \times 20 \times 20$ **k**-point mesh for integrals over the Brillouin zone, and a 24 point semi-

**Table 4.1** *Hutsepot*-optimised lattice parameters for each of the pure refractory elements (V, Nb, Mo, Ta, W) with their experimentally determined values for comparison. As expected, the LDA is found to slightly underestimate lattice spacings compared to their experimental values

| Element | $a_{DFT}$ (Å) | $a_{Exp.}$ (Å) |
|---|---|---|
| V | 2.904 | 3.039 |
| Nb | 3.222 | 3.301 |
| Mo | 3.086 | 3.147 |
| Ta | 3.225 | 3.303 |
| W | 3.106 | 3.158 |

circular Gauss-Legendre grid in the complex plane to integrate over valence energies. We use the LDA and the exchange-correlation functional of Perdew-Wang [28].

Along with the three multicomponent systems mentioned above, here we choose to study every possible equiatomic binary subsystem of the five-component VNbMoTaW,[1] with the aim of understanding the underlying physics driving ordering in the multicomponent systems.

To demonstrate that this approach need not rely on experimentally determined lattice spacings, rather than taking experimental values for these systems, we instead choose to obtain all lattice parameters *ab initio*, finding the value for which the total DFT energy at zero temperature is minimised. This is achieved by performing self-consistent calculations for a range of lattice spacings, then fitting a quadratic to the total energy and obtaining the lattice spacing minimising the energy. In principle, one could consider the effects of thermal expansion of the lattice with increasing temperature, which may slightly alter the predictions of our subsequent linear response calculation. We have not done so in this chapter, and we expect the effects for the metallic systems considered here to be comparatively small. This aspect will be addressed explicitly in Chap. 5.

First we perform calculations for the pure elements, for which results are tabulated in Table 4.1. Using *Hutsepot* and the LDA, we find a small but systematic underestimation of lattice parameter compared to experimental values, which are taken from [8]. This is to be expected; the LDA is known to underestimate lattice parameters compared to experiment [29]. We then optimise lattice parameters for the considered alloys, results of which are given in Table 4.2, where they are also compared to the predictions of Vegard's law. Vegard's law is an empirical finding which states that the lattice parameter of a solid solution of two or more elements is approximately the weighted mean of the separate values for the two pure elements [30, 31]. We find no significant discrepancies between the Vegard predictions and the DFT-optimised values.

Having optimised the lattice parameters, we then visualise the density of states for each system. An indicative subset of DoS plots is shown in Fig. 4.1, while plots

[1] There are 12 possible binaries in total: VNb, VMo, VTa, VW, NbMo, NbTa, NbW, MoTa, MoW, and TaW.

**Table 4.2** *Hutsepot*-optimised lattice parameters for the considered refractory alloys compared to the prediction of Vegard's law based on the optimised lattice parameters for the pure elements in Table 4.1. For the Vegard's law prediction, we take a weighted average of the *Hutsepot*-optimised lattice parameters for the pure elements

| Material | $a_{\mathrm{DFT}}$ (Å) | $a_{\mathrm{Vegard}}$ (Å) |
|---|---|---|
| VNb | 3.093 | 3.063 |
| VMo | 3.011 | 2.995 |
| VTa | 3.105 | 3.065 |
| VW | 3.035 | 3.005 |
| NbMo | 3.147 | 3.154 |
| NbTa | 3.226 | 3.224 |
| NbW | 3.156 | 3.164 |
| MoTa | 3.146 | 3.156 |
| MoW | 3.098 | 3.096 |
| TaW | 3.157 | 3.166 |
| NbMoTa | 3.173 | 3.178 |
| NbMoTaW | 3.151 | 3.109 |
| VNbMoTaW | 3.119 | 3.160 |

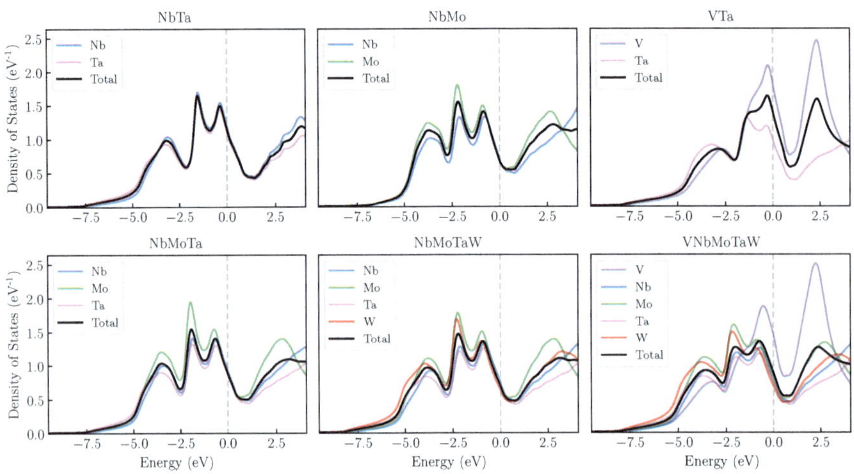


**Fig. 4.1** Comparison of the total and species-resolved DoS for the refractory HEAs considered in this work, with three sample equiatomic binary compositions for comparison. The only 3*d* element, V, can be seen to have a narrower *d* band than the 4*d* and 5*d* elements. Figure reproduced from [29] with permission from the American Physical Society

for all considered binary systems and the pure elements are shown in Fig. 4.2. These plots can be understood by considering the nature of the partially filled *d*-electron states for each element. For a brief introduction to the electronic structure of metals and alloys, see Chap. 3, Sect. 3.9.

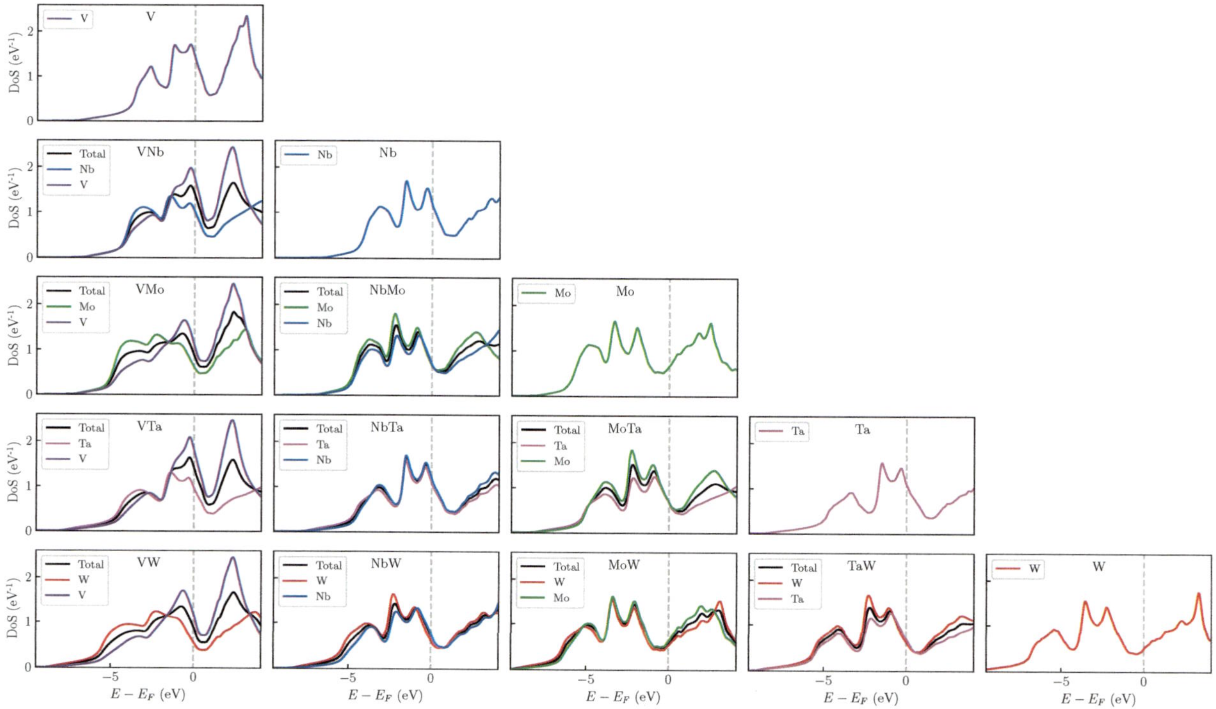


**Fig. 4.2** Plots of the total and species-resolved density of states curves for the five pure refractory elements, as well as all ten possible equiatomic binaries. All are modelled on the bcc lattice, with lattice parameters from Tables 4.1 and 4.2. Figure reproduced and adapted from [29] with permission from the American Physical Society

The only 3*d* metal here, V, has by far the narrowest *d*-band, and later we will see that this narrow *d*-band, combined with a notable discrepancy in atomic sizes, drives B32-like order for alloys containing V, which can be understood in a similar manner to the mechanisms driving ordering in the Ni-Pt system [32].

The bandwidth difference can be most clearly seen in Fig. 4.1, when comparing the DoS plot for NbTa with that for VTa. The tail of the *d*-band can be seen to have been pulled to lower energies in VTa than in NbTa, indicative of hybridised bonding states forming at lower energies. In addition, the width of the *d*-band associated with V can be seen, overall, to be narrower than that of Ta. Thirdly, there is significant difference in the overall character of the species-resolved DoS curve associated with V compared to that of Nb or Ta. We intuitively expect all of these features to drive stronger correlations between the 3*d* element V and the other 4*d*/5*d* elements. Finally, although not immediately apparent from the DoS plots, in all self-consistent calculations there is significant transfer of charge to V from the heavier elements. This is a consequence of the relative atomic size difference and the lattice parameter of the system lying between those of the pure elements. Charge from the Wigner-Seitz cell of larger atoms such as Nb and Ta effectively 'spills over' into the Wigner Seitz cell associated with V. This atomic size difference, manifested as a charge transfer in our calculations, is expected to drive strong atom-atom correlations in V-containing systems.

The species-resolved curves for Nb, Mo, Ta, and W are instead understood by considering valence differences, as the effects of differing atomic size and band-width are no longer as significant between these elements. 4*d*/5*d* pairs such as Nb/Ta are isoelectronic, so their *d*-bands are very similar in character around $E_{\mathrm{F}}$, best illustrated in the curves for NbTa and MoW in Fig. 4.2. This similarity leads to these elements forming stable solid solutions down to low temperatures, with little to no ASRO observed. Pairs of 4*d*/4*d*, 4*d*/5*d* or 5*d*/5*d* elements with a valence difference can be seen to shift the position of their *d* bands to produce a curve with a profile somewhere between the profiles of the DoS curves for the two pure elements, best illustrated by the curves for pure Mo, pure Ta, and the binary MoTa in Fig. 4.2. Although there is no significant charge transfer between these elements, the hybridisation and adjustment of the partially filled *d*-bands is expected to produce stronger atom-atom correlations than in isoelectronic systems. It should be pointed out that the above observations are consistent with the empirically determined Hume-Rothery rules predicting whether or not two metallic elements will alloy and form a solid solution [33].

The mechanisms discussed above can be seen to be in play in the three multi-component systems in Fig. 4.1, where the standout element is again V in the five-component VNbMoTaW. The valence differences between elements such as Ta/W and Nb/Mo, manifested in relative shifts in energy between species-resolved curves, can also be clearly seen.

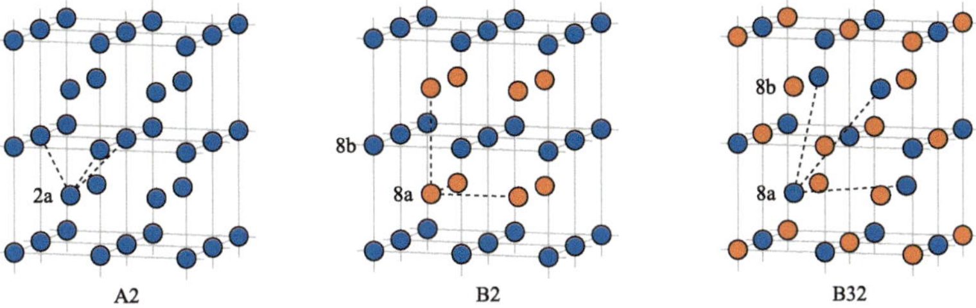


**Fig. 4.3** Visualisations of the relevant ordered structures imposed on the bcc lattice for the refractory alloys of this chapter. All three structures can be described by concentration waves. A2 represents a disordered alloy. B2 can be described by modes $\mathbf{k} = (0, 0, 1)$ and equivalent, while B32 is described by modes $\mathbf{k} = (\frac{1}{2}, \frac{1}{2}, \frac{1}{2})$. Figure reproduced from [29] with permission from the American Physical Society

## 4.3 Linear Response

Proceeding, we now use the linear response theory outlined in Chaps. 2 and 3 calculate $S^{(2)}_{\alpha\alpha'}(\mathbf{k})$ and, subsequently, to construct the chemical stability matrix in reciprocal space for these systems. Visualised in Fig. 4.4a are plots of $S^{(2)}_{\alpha\alpha'}(\mathbf{k})$ around high-symmetry lines of the irreducible Brillouin zone for the same six systems as in Fig. 4.1. Then, in Fig. 4.4b, we construct the chemical stability matrix, $[\beta^{-1}\Psi^{-1}(\mathbf{k})]_{\alpha\alpha'} = \left[\beta^{-1}\frac{\delta_{\alpha\alpha'}}{c_\alpha} - S^{(2)}_{\alpha\alpha'}(\mathbf{k})\right]$, at $T = 1200$ K and visualise its eigenvalues. In addition, we visualise eigenvalues of the chemical stability matrix for all ten possible equiatomic binaries in Fig. 4.5, again, all at a temperature of $T = 1200$ K. (Recall that, for an $s$-component alloy, the chemical stability matrix has $s - 1$ physically meaningful eigenvectors.) Then, in Table 4.4, we provide the associated eigenvectors for the multicomponent systems at various high-symmetry points of the Brillouin Zone. In Table 4.3, we give predicted temperatures, wavevectors, and chemical polarisations describing disorder-order transitions in these systems. Finally, in Fig. 4.6, we visualise examples of the predicted partially ordered structures for NbMoTaW and VNbMoTaW.

$S^{(2)}_{\alpha\alpha'}(\mathbf{k})$ can be thought of as the lattice Fourier transform of an effective pair interaction, so visualisation of this data provides information about the range and strength of chemical interactions for a given system. Then, the chemical stability matrix provides information about dominant chemical orderings. We see that interactions are typically strongest (indicated by large $S^{(2)}_{\alpha\alpha'}(\mathbf{k})$) between V and other considered elements, while notable interactions are also present for pairs with a valence difference, such as Nb-Mo and Ta-W. Pairs of $4d/5d$ elements which are isoelectronic, such as Nb-Ta, have very flat associated $S^{(2)}$ data.

Proceeding, we examine the eigenvalues of the chemical stability matrices in reciprocal space. To understand the multicomponent systems, we first look to the three example binary systems in Fig. 4.4b. The isoelectronic $4d/5d$ binary, NbTa, has one exceptionally flat concentration wave mode, associated with weak interac-

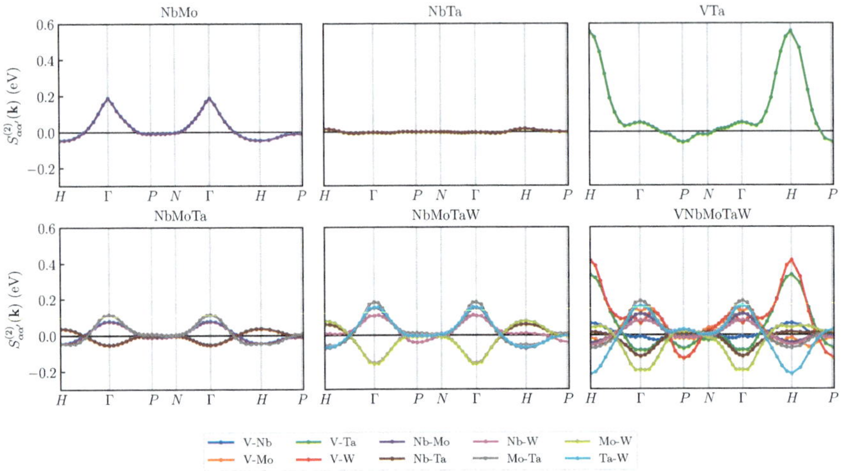


(a) $S^{(2)}_{\alpha\alpha'}(\mathbf{k})$

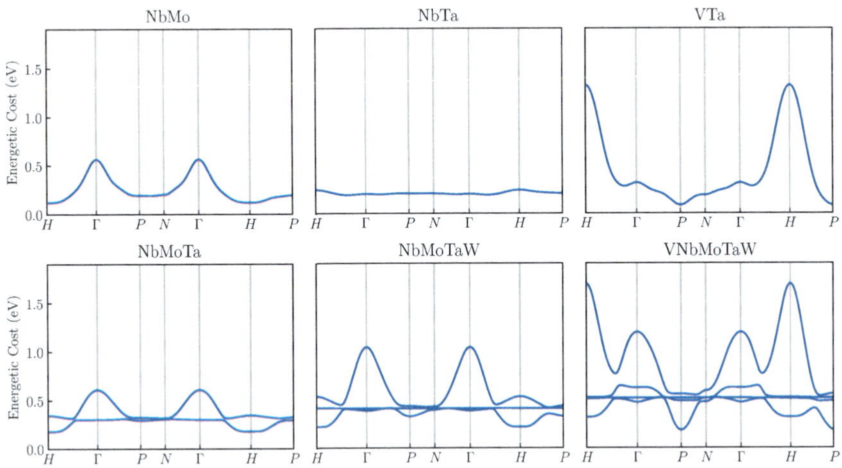


(b) Eigenvalues of $[\beta^{-1}\Psi^{-1}(\mathbf{k})]_{\alpha\alpha'} = \left[\beta^{-1}\frac{\delta_{\alpha\alpha'}}{c_\alpha} - S^{(2)}_{\alpha\alpha'}(\mathbf{k})\right]$ at $T = 1200$ K

**Fig. 4.4** Plots of the raw $S^{(2)}_{\alpha\alpha'}(\mathbf{k})$ data (top) and eigenvalues of the chemical stability matrix (bottom), around the irreducible Brillouin Zone of the body-centred cubic lattice for the multicomponent refractory alloys, with three sample equiatomic binary compositions for comparison. The chemical stability matrix is evaluated at $T = 1200$ K. The $S^{(2)}_{\alpha\alpha'}(\mathbf{k})$ includes the Onsager correction. A minimum eigenvalue at $H$ indicates B2-like ordering is preferred, while a minimum at $P$ indicates a preference for B32-like order. The bottom pane of the figure is reproduced from [29] with permission from the American Physical Society

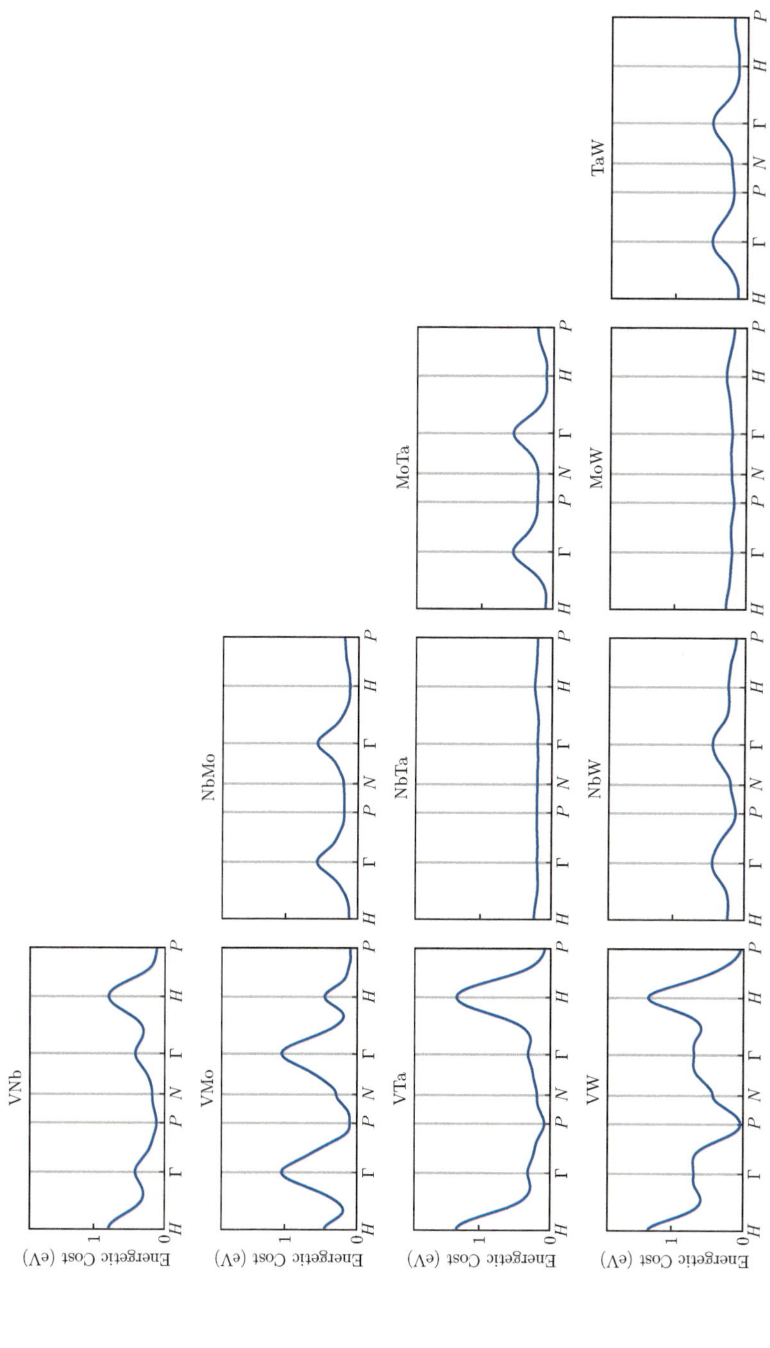


**Fig. 4.5** Plots of the eigenvalues of the chemical stability matrix for all possible equiatomic binary alloys constructed from the five refractory metals: V, Nb, Mo, Ta, and W. It can be seen that the inclusion of V drives B32-like ordering, indicated by the minima at $P$. Figure reproduced and adapted from [29] with permission from the American Physical Society

**Table 4.3** Transition temperatures, wavevectors, and chemical polarisations for the considered refractory systems, together with the classification of the indicated ordered structure. The chemical polarisation is the eigenvector in composition space associated with the eigenvalue passing through zero

| Material | $T_{us}$ (K) | $\mathbf{k}_{us}$ ($\frac{2\pi}{a}$) | Structure | $\Delta c_V$ | $\Delta c_{Nb}$ | $\Delta c_{Mo}$ | $\Delta c_{Ta}$ | $\Delta c_W$ |
|---|---|---|---|---|---|---|---|---|
| VNb | 504 | $(\frac{1}{2}, \frac{1}{2}, \frac{1}{2})$ | B32 | 0.707 | −0.707 | | | |
| VMo | 593 | $(\frac{1}{2}, \frac{1}{2}, \frac{1}{2})$ | B32 | 0.707 | | −0.707 | | |
| VTa | 691 | $(\frac{1}{2}, \frac{1}{2}, \frac{1}{2})$ | B32 | 0.707 | | | −0.707 | |
| VW | 905 | $(\frac{1}{2}, \frac{1}{2}, \frac{1}{2})$ | B32 | 0.707 | | | | −0.707 |
| NbMo | 513 | (0, 0, 1) | B2 | | 0.707 | −0.707 | | |
| NbTa | 86 | (0, 0, 0.4) | N/A | | 0.707 | | −0.707 | |
| NbW | 462 | $(\frac{1}{2}, \frac{1}{2}, \frac{1}{2})$ | B32 | | 0.707 | | | −0.707 |
| MoTa | 602 | (0, 0, 1) | B2 | | | 0.707 | −0.707 | |
| MoW | 242 | $(\frac{1}{2}, \frac{1}{2}, \frac{1}{2})$ | B32 | | | 0.707 | | −0.707 |
| TaW | 488 | (0, 0, 1) | B2 | | | | 0.707 | −0.707 |
| NbMoTa | 511 | (0, 0, 1) | B2 | | −0.406 | 0.816 | −0.410 | |
| NbMoTaW | 559 | (0, 0, 1) | B2 | | −0.383 | 0.594 | −0.595 | 0.383 |
| VNbMoTaW | 742 | $(\frac{1}{2}, \frac{1}{2}, \frac{1}{2})$ | B32 | −0.824 | 0.012 | 0.085 | 0.252 | 0.500 |

tions between species and, consequently, weak ordering tendencies and a low atomic ordering temperature. The $4d/4d$ binary with a valence difference, NbMo, has a concentration wave mode which peaks at $\Gamma$ and has a minimum at $H$, which is indicative of B2-like ordering tendencies.[2] Finally, the $3d/5d$ system with a large atomic size difference, VTa, has a strongly varying concentration wave mode, associated with strong interactions between these species. This mode, dipping at $P$ and peaking at $H$, is indicative of B32-like ordering. The A2, B2, and B32 structures are visualised in Fig. 4.3 for reference.

We understand this B32-like ordering to have its origins in an effective charge transfer from the large atoms (Ta) to small atoms (V) as discussed in Sect. 4.2, screened by the valence electrons. The argument that the screened charge transfer drives B32-like order is as follows. Given a Coulombic interaction limited to the first two shells,[3] and an 'alloy' of positive and negative charges, the B32 structure is energetically favoured over the B2 structure. This is because in the B2 arrangement each charge has 8 unlike and 6 like neighbours in the first two coordination shells, while in the B32 structure it is 10 unlike and 4 like. The Coulombic interaction, which favours unlike pairs of charges, therefore preferentially selects the B32 ordering.

---

[2] Recall that in Chap. 2 we saw that the B2 structure can be described by a concentration wave with $\mathbf{k}_{us} = (0, 0, 1)$ and equivalent.

[3] This is a reasonable suggestion for the bcc structure, where first- and second-nearest neighbour shells lie at very similar radial distances to one another.

**Table 4.4** Eigenvalues of the chemical stability matrix for the considered multicomponent refractory alloys at selected special points of the IBZ. These are evaluated at $T = 1200$ K and their corresponding eigenvectors, or polarisations, in concentration space are also provided. By inspecting Fig. 4.4b it is possible to use these tabulated values to understand the chemical fluctuations to which each visualised concentration wave mode corresponds

| Material | Point | Eigenvalue (eV) | $\Delta c_1$ | $\Delta c_2$ | $\Delta c_3$ | $\Delta c_4$ | $\Delta c_5$ |
|---|---|---|---|---|---|---|---|
| NbMoTa | $H$ | 0.178 | −0.408 | 0.816 | −0.410 | | |
| | | 0.346 | −0.708 | 0.002 | 0.706 | | |
| | Γ | 0.303 | 0.745 | −0.083 | −0.662 | | |
| | | 0.605 | 0.334 | −0.812 | 0.478 | | |
| | $P$ | 0.294 | −0.800 | 0.543 | 0.256 | | |
| | | 0.324 | −0.166 | −0.610 | 0.775 | | |
| NbMoTaW | $H$ | 0.220 | −0.383 | 0.594 | −0.595 | 0.383 | |
| | | 0.405 | 0.599 | −0.423 | 0.529 | −0.456 | |
| | | 0.535 | 0.494 | 0.467 | −0.286 | −0.675 | |
| | Γ | 0.388 | 0.516 | 0.474 | −0.399 | −0.591 | |
| | | 0.414 | −0.541 | 0.468 | 0.529 | −0.456 | |
| | | 1.023 | 0.436 | −0.554 | 0.557 | −0.440 | |
| | $P$ | 0.333 | 0.636 | 0.087 | 0.042 | −0.765 | |
| | | 0.412 | 0.581 | −0.614 | −0.360 | 0.394 | |
| | | 0.437 | −0.086 | −0.604 | 0.786 | −0.097 | |
| VNbMoTaW | $H$ | 0.318 | 0.079 | 0.328 | −0.638 | 0.591 | −0.360 |
| | | 0.510 | −0.254 | 0.064 | 0.625 | 0.255 | −0.690 |
| | | 0.525 | 0.191 | −0.827 | 0.036 | 0.521 | 0.079 |
| | | 1.680 | −0.832 | 0.062 | −0.003 | 0.339 | 0.435 |
| | Γ | 0.472 | 0.307 | 0.434 | 0.349 | −0.521 | −0.569 |
| | | 0.517 | 0.013 | 0.487 | −0.520 | −0.487 | 0.506 |
| | | 0.615 | −0.792 | 0.535 | 0.159 | 0.218 | −0.120 |
| | | 1.173 | 0.281 | 0.297 | −0.619 | 0.494 | −0.454 |
| | $P$ | 0.197 | −0.824 | −0.012 | 0.085 | 0.252 | 0.500 |
| | | 0.488 | 0.286 | −0.704 | 0.122 | −0.279 | 0.575 |
| | | 0.517 | −0.059 | −0.512 | 0.510 | 0.516 | −0.456 |
| | | 0.561 | −0.188 | 0.206 | 0.720 | −0.626 | −0.112 |

We find that this effect dominates in all the considered binaries containing V, dominating any valence differences, as visualised in Fig. 4.5. This effect is strongest in VW, a known B32-former [34], where we predict an atomic ordering temperature of 905 K, but is also present in all the other V-containing binaries, for which there is limited existing literature with which to compare.

We look to the above discussion of the considered binary systems to understand the concentration wave modes in the multicomponent systems visualised in Fig. 4.4b. It can be clearly seen that modes associated with the discussed mechanisms are present in these multicomponent systems, albeit with mixing between modes. Sample eigenvalues and eigenvectors are tabulated in Table 4.4. The ternary NbMoTa has two concentration wave modes. The first, near flat mode is associated with interactions between isoelectronic elements and, in concentration space, is polarised along the Nb and Ta 'directions.' The second mode peaks at $\Gamma$ and has a minimum at $H$, and is associated with the valence difference and a B2-like ordering, with a chemical polarisation associated with an increase of Mo and a relative (and almost equal) decrease in Nb and Ta. These two modes can also be seen in the four component NbMoTaW, although now we begin to pick up traces of atomic size differences, primarily between W and Nb, which was detected in the NbW binary and shows up in the small local minimum eigenvalue at $P$. Inspecting the chemical polarisations of Table 4.4 clarifies the interpretation of each mode. Finally, for the five-component VNbMoTaW, we see it is the atomic size difference that dominates, with a concentration wave mode dipping at $P$ and peaking at $H$ associated with the relative size difference between V and the other elements. We also see traces of the isoelectronic and valence-difference modes discussed for the systems without V present, but the energy scale of these is not as large as for the V-driven mode, leading to a predicted B32-like ordering.

The predicted ordering temperatures are broadly consistent with earlier works. For NbMoTaW, our prediction of a B2-like ordering (visualised in Fig. 4.6) is in good agreement with a number of earlier works [16–22, 35]. The predicted temperature of 559 K for this ordering is also in good agreement with some earlier works. Körmann and Sluiter [18] used the projector-augmented wave (PAW) method with chemical disorder simulated using special quasi-random structures (SQS) and included effects of vibrations, electronic excitations, and configurational entropy. These authors found a transition temperature of 717 K without lattice relaxations, while with lattice lattice relaxations the transition temperature was reduced to 508 K. More recently, [35] used interatomic potentials generated via machine learning to predict an ordering temperature to a B2 phase of approximately 600 K without atomic relaxations, and

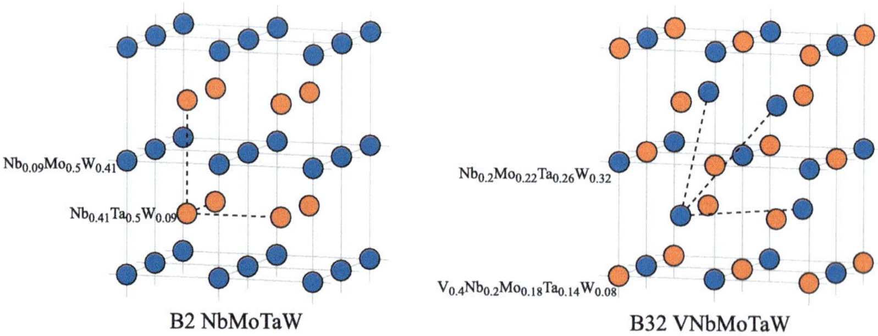


**Fig. 4.6** Visualisations of the partially ordered structures predicted by the Landau-type linear-response theory for NbMoTaW (left) and VNbMoTaW (right), emerging at 559 and 742 K respectively. The partial occupancies of the two non-equivalent sites in each crystal structure are specified by the labels on the figures

approximately 300 K when their effects are included. We note that direct inclusion of the effects of lattice relaxations is not currently possible within the KKR formulation of DFT as used in this work, so the agreement of our results with those for unrelaxed systems is pleasing.

In the five-component VNbMoTaW, both the nature of atomic order (B32-like) and the predicted ordering temperature (742K) are in superb agreement with [10], which used a cluster expansion Hamiltonian to predict B32-like order emerging at around 750 K. That our results agree with the results of a study using a cluster expansion is reassuring; our pairwise Hamiltonian represents a cluster expansion restricted to pairwise interactions only. Future work could investigate the significance of higher-order terms and their effect on model predictions. The B32 ordering in VNbMoTaW, and the same predicted ordering in the VW binary, are also consistent with analyses of the V-W binary system, in which a B32 ordering is predicted [34, 36]. We note that our predicted B32 ordering temperature (905 K) for the binary VW system is higher than that predicted in [36] ($\sim$330 K), and attribute this to the fact that our modelling does not account for local lattice distortions [35] or thermal expansion of the lattice. In addition, our ordering temperature is obtained within a mean-field theory, which naturally results in an overestimate of predicted ordering temperatures.

### 4.3.1 The Importance of the Onsager Correction

Recall that the Onsager correction detailed in Chaps. 2 and 3 served to restore on-site sum rules, including those concerning the on-site charge response. For these systems, particularly those containing V, where there is a large inherent charge transfer in the self-consistent DFT calculations, we find the Onsager correction to be significant. Visualised in Fig. 4.7 are plots of the eigenvalues of the chemical stability matrix for selected refractory systems evaluated at $T = 1200$ K with the appropriate Onsager correction, compared with results when the Onsager correction is evaluated at high temperature, where its effects are negligible.

For an isoelectronic system with negligible charge transfer, such as NbTa, the effect of the correction is insignificant. For systems with a valence difference and a small degree of associated charge transfer, such as NbMo, the correction is seen to have a small, but visible effect. For systems containing V, however, the effect is significant; VTa can be seen to have already been predicted to go through a disorder-order transition at this temperature, while VNbMoTaW is seen to have a significantly lower-lying minimum eigenvalue, and an associated increased predicted ordering temperature. From these results, we stress that the Onsager correction is an important and necessary aspect of the modelling approach, particularly for systems in which the effects of charge transfer between chemical species are large.

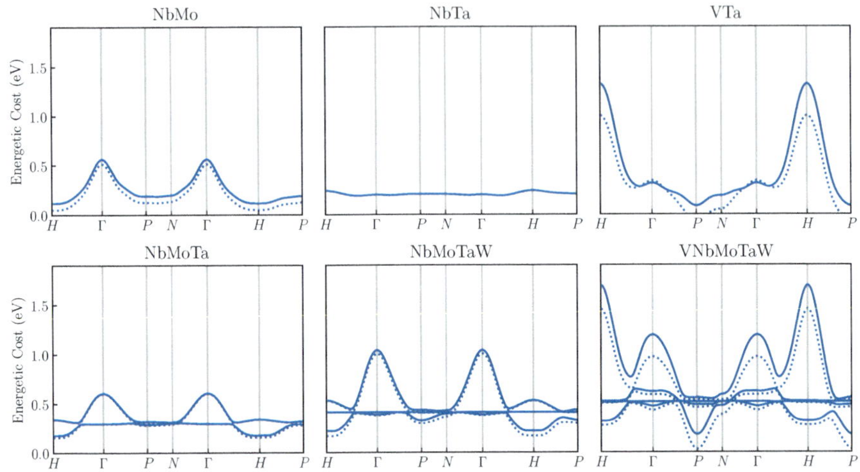


**Fig. 4.7** Plots of the eigenvalues of the chemical stability matrix for selected refractory systems evaluated at $T = 1200$ K, with dashed lines showing results when the Onsager correction is evaluated at high temperature where its effects are negligible. The Onsager correction, which restores on-site sum rules, lowers the predicted transition temperature significantly for systems containing V by restoring sum rules relevant to the charge response

## 4.3.2 The Failure of the Band-Only Approximation

In addition to the Onsager correction mentioned above, we also stress the importance of including fully the effects of charge transfer and charge response, as outlined in Chap. 3. Visualised in Fig. 4.8 are the eigenvalues of the chemical stability matrix computed within the band-only approximation, with eigenvalues computed using the full $S^{(2)}$ theory shown for comparison, both evaluated at $T = 2000$ K. In NbMoTa and NbMoTaW, the minimum at the $H$ point is seen to be broadened and we find the minimum no longer passes through the $H$ point exactly. This is indicative of

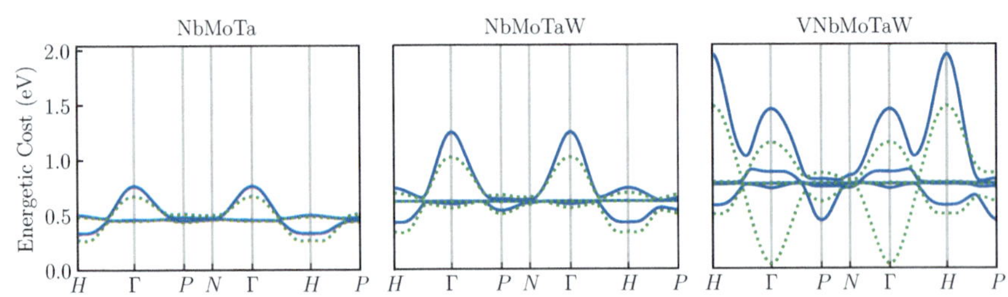


**Fig. 4.8** Plots of the eigenvalues of the chemical stability matrix for selected refractory systems evaluated at $T = 2000$ K, with dashed green lines showing results computed using the band-only approximation, and solid blue lines showing results from the full $S^{(2)}$ theory. The band-only approximation is particularly poor for the five-component VNbMoTaW

long-ranged interactions, as suggested in some earlier work on this system [19, 20]. Once the effects of charge-transfer and charge-response are included, however, the minimum at $H$ becomes much more well-defined, and a simple B2 ordering is predicted.

For the five-component VNbMoTaW, the situation is far worse. The band only approximation erroneously predicts that V phase segregates out of the solid solution at approximately 1700 K, a result at odds with experimental evidence [8, 9]. Again, fully including the effects of charge-transfer and charge-response remedies this, and V is found to integrate well into the solid solution. We therefore stress that any approach for modelling atomic arrangements in multicomponent alloys based on the CPA should take care to ensure that these effects are well-accounted for; the band-only approximation may prove insufficient in a number of alloy systems.

## 4.4 Pairwise Interactions

Starting from the chemical stability matrix evaluated in reciprocal space, it is possible to backwards Fourier transform to recover a pairwise, real-space interaction as specified in Chap. 2. For these systems, we visualise the pairwise interactions as a function of distance for a fit to the first 12 coordination shells in Fig. 4.9. Interactions are clearly dominated by the first two coordination shells. However, it can be seen that particularly for interactions associated with V, such as V-V, V-Ta, and V-W, that the interaction has a long 'tail.' This is associated with the charge transfer between elements and the inherent long-range of the Coulomb interaction. However, as these systems are metallic, there is a degree of electronic screening, meaning that interactions are still negligible beyond the first five or six coordination shells. For the purposes of the lattice-based atomistic modelling, we truncate our fit to to the first six coordination shells of the bcc lattice, and find that this still captures the reciprocal-space data with good accuracy.

Appendix C, Tables C.1, C.2, and C.3 then explicitly tabulate our fitted $V^{(n)}_{\alpha\alpha'}$ for $n = 1, 2, 3, 4, 5, 6$ for the three multicomponent systems for reference. Given that interactions on the second coordination shell are sometimes larger than those on the first, we conclude that any model limited purely to nearest-neighbour interactions will fail to capture the relevant physics. Interactions being strongest on the first two neighbour shells is also consistent with recent results obtained using effective pair interactions for NbMoTaW and VNbMoTaW generated using machine learning on a DFT data-set [37], which found strong interactions on the first two coordination shells, but no significant long-ranged interactions.

Inspecting Fig. 4.9 and Tables C.1, C.2, and C.3, it can be seen that pairs of $4d/5d$ elements with a valence difference interact favourably on the first coordination shell, while $4d/5d$ pairs which are isoelectronic are less favoured, consistent with the B2-like ordering predicted in the linear response analysis. The outlier is the only $3d$ element considered here, V, which interacts significantly more strongly, both with

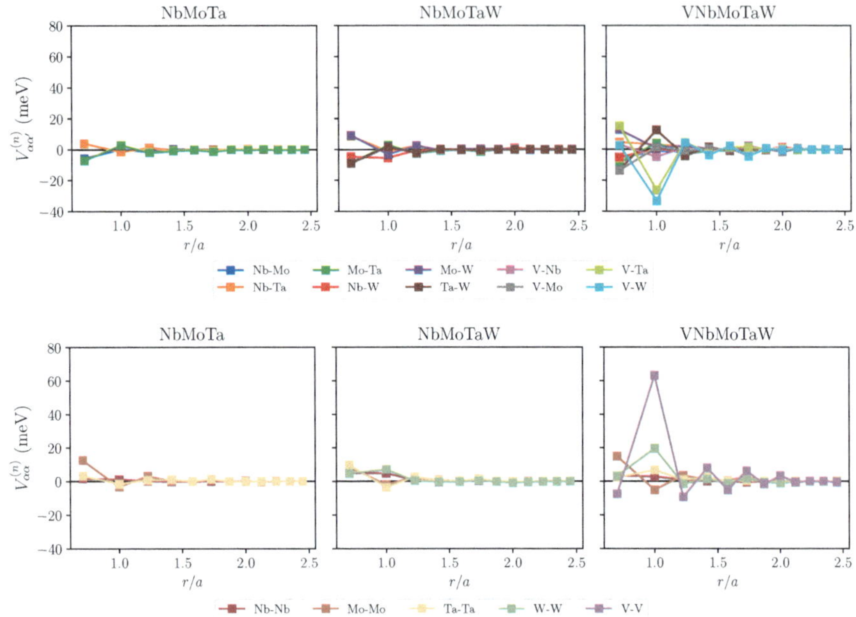


**Fig. 4.9** Visualisation of $V_{\alpha\alpha'}^{(n)}$ for the three multicomponent refractory systems considered, fitted to the first twelve coordination shells. It can be seen that interactions are dominated by the first two coordination shells

itself and with the other elements present, on the second shell than the first, consistent once again with B32-like order.

## 4.5 Atomistic Modelling

Using the obtained pairwise interactions, we performed lattice-based Monte Carlo (MC) simulations for NbMoTa, NbMoTaW, and VNbMoTaW to better understand the nature of ASRO in these systems and probe ordering below the initial disorder-order transition temperature. A lattice-based model is suitable for these systems because the BCC structure is capable of accommodating large atomic size discrepancies [39]. All calculations used a system of $16 \times 16 \times 16$ cubic unit cells, each with 2 lattice sites per unit cell, for a total of 8192 atoms in the simulation cell, with periodic boundary conditions applied. That this system size was appropriate was verified by conducting simulations on two smaller systems ($8 \times 8 \times 8$ and $4 \times 4 \times 4$ unit cells), and it was observed that that the quantities of interest were adequately converged. The systems were prepared in an initially random configuration, then annealed from 1200 to 10 K in steps of 1 K, with $10^4$ MC steps per atom at each temperature. A burn-in period was applied to the simulation at each temperature to ensure the system was equilibrated at each temperature before statistics were drawn. An ensemble of 32 simulations was used for each alloy composition and the results averaged. This enables extraction of an error measurement on quantities such as the specific heat capacity of the simulation, quantifying the uncertainty on the results.

Figure 4.10 shows plots of the Warren-Cowley ASRO parameters and SHC curves for NbMoTa, NbMoTaW, and VNbMoTaW, while Fig. 4.11 shows sample visualised configurations from our simulations. The images visualising the configurations are generated using OVITO [38].

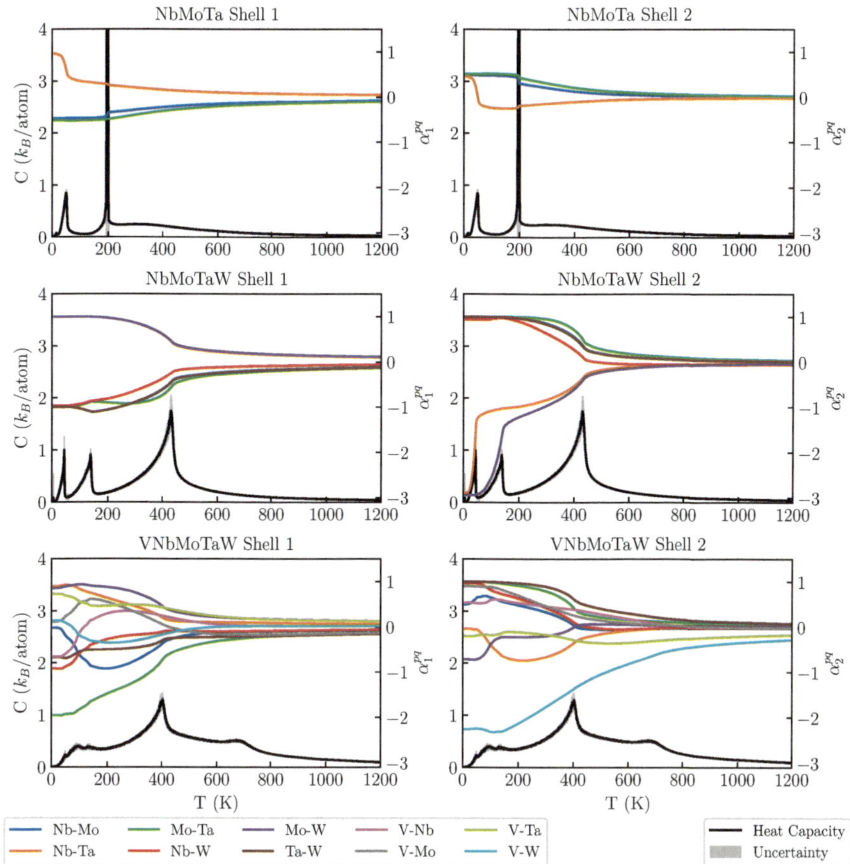


**Fig. 4.10** Plots of the Warren-Cowley ASRO parameters and configurational contribution to the SHC as a function of temperature for the three multicomponent systems considered from lattice-based Monte Carlo simulations using our extracted pairwise parameters. We show $\alpha_n^{pq}$ for $n = 1, 2$. NbMoTa shows little ASRO, with the only notable feature being towards pairs with a valence difference on the first shell, and away from those pairs on the second shell, indicative of B2-like ordering. The same trend can be seen more strongly in NbMoTaW. The most notable trend on the five component plots is that V-W is largely indifferent on the first shell, but highly favoured on the second, a precursor to a B32-like structure

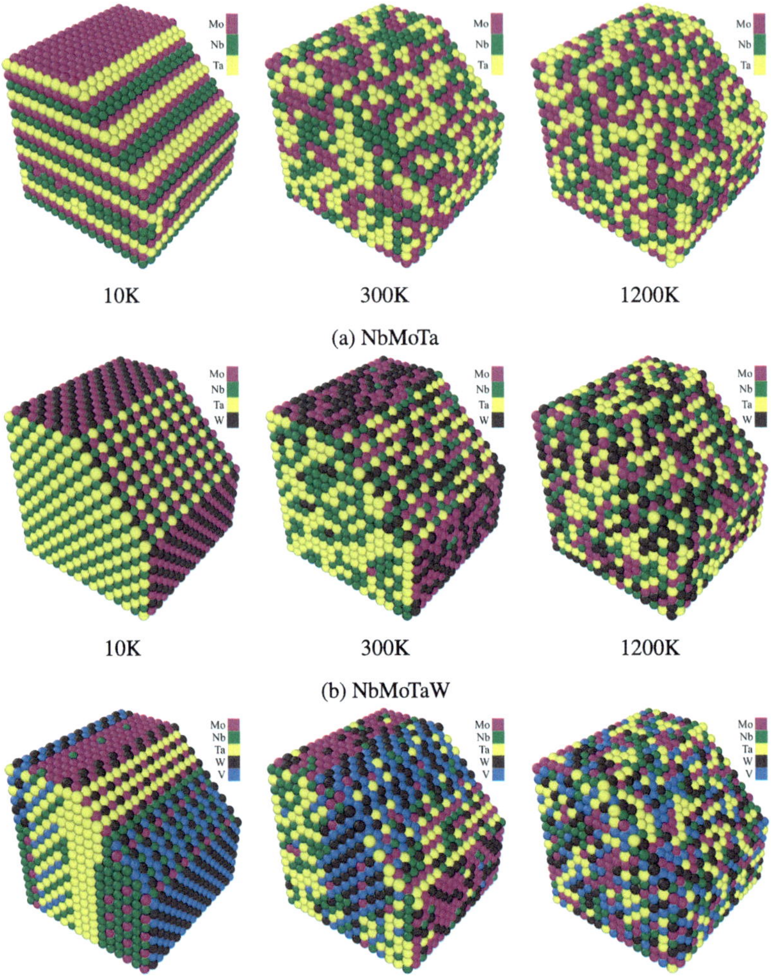


**Fig. 4.11** Visualised configurations from Monte Carlo simulations for the three considered multicomponent refractory systems at temperatures of 1200, 300, and 10 K. V, Nb, Mo, Ta, and W are coloured blue, green, pink, yellow, and black, respectively. A cut has been made through the simulation cell to make ordered structures more clearly visible. In the case of NbMoTaW, the emergence of a layered, B2-like structure can be seen in the $T = 300$ K configuration. For the ternary NbMoTa, some layering can be seen relative to the 1200 K configuration, but it is less clear than for the quarternary. For the five component system, the system demonstrates multiphase behaviour, with patches of B32-like order between V and W, and other patches of order akin to that observed in the ternary NbMoTa. Images generated using Ovito [38]

In NbMoTa, little ASRO is observed until very low temperatures, although it is consistently seen that Nb and Ta favour Mo as a nearest neighbour, indicative of a B2-like ordering and consistent with our earlier linear response analysis. The visualised low-temperature configuration suggests the possibility of an intriguing, layered ground state, although this emerges at such low temperatures as to be purely speculative rather than experimentally relevant. This layering is reflective of the preference for nearest-neighbour pairs with a valence difference. The two peaks in the specific heat capacity are understood to originate (in order of temperature at which they occur from highest to lowest) from a B2-like ordering between Mo and Nb/Ta, then a subsequent low-temperature ordering between Nb and Ta. We note that these low temperature results differ slightly from those reported in [29], which fitted interactions to the first four coordination shells of the bcc lattice. This difference suggests that these weak interactions on more distant coordination shells play a subtle role in determining the ground-state configuration of the alloy, even if they are unimportant at experimentally-relevant temperatures.

For the well-studied four component NbMoTaW, it can be clearly seen that Mo-W and Nb-Ta pairs (isoelectronic) are disfavoured as nearest neighbours, while pairs Nb-W, Mo-Ta, Nb-Mo, and Ta-W (pairs with a valence difference) are favoured, indicative of B2-like order, consistent again with our linear response analysis and also with earlier literature. This B2-like ordering is followed at lower temperatures by ordering on each of the two sublattices, resulting in a Heusler-type ground state, as can be seen in the visualised configurations. The ordering between Mo and W on one sublattice emerges earlier than between Nb and Ta in our simulations. The ground state obtained in our simulations is not the same as in some other works, notably [35], which predicted a layered arrangement. However, our approach is most valid at high temperatures, where the pairwise parameters for atomistic modelling are extracted, so we do not view this low-temperature disagreement as an issue.

Finally, for the five component VNbMoTaW, the picture is less clear-cut, and a number of separate peaks in the heat capacity curve can be seen. The strongest trends are towards Mo-Ta and Nb-Mo pairs at nearest neighbour distance, which could be interpreted as B2-like ordering, as for the four component system. However, at second-nearest neighbour distance, V-W pairs are highly favoured, suggesting instead B32-like order. When we visualise our configurations in Fig. 4.11 the picture becomes clear, however, because there is clear multiphase behaviour emerging, with separate regions of B32-like V-W and patches of B2-like order involving the other elements present, although this emerges at low temperatures and is unlikely to be experimentally observable. These results emphasise the importance of the effective pair interactions in determining the phase behaviour of a given system below and initial ordering inferred via the Landau theory.

## 4.6 Computational Cost

Before summarising the results of this chapter, it is perhaps pertinent to assess the computational (and, therefore, environmental) cost of performing these calculations. At present, the environmental impact of high-performance computing is under increasing scrutiny [40], and DFT-based approaches are not famed for being computationally cheap. A conventional single-point energy DFT calculation on *one* 128 atom supercell could we would expect to take roughly (order of magnitude) 100 core hours on a Warwick computing cluster, and 10s or 100s of these evaluations would be needed to train a cluster expansion or machine-learned interatomic potential for subsequent study.

By comparison, a quick, back of the envelope calculation estimates that *all* data tabulated or visualised in this chapter can be reproduced in less than 1000 core hours on the Avon cluster at the University of Warwick, which uses Intel Xeon Platinum 8268 (Cascade Lake) processors. Self-consistent potentials and DoS calculations represent less than 5 core hours per material. The linear response code developed in house takes a little longer, at around 40 core hours per material to generate the data necessary for eigenvalue plots, transition temperatures/modes, and atom-atom pair interactions, but we emphasise here that the main portion of the linear response calculations does not scale with the number of components considered. Finally, our atomistic model for MC simulations makes these calculations exceptionally cheap, and each simulation runs on a single core for no more than 24 h. (Although averaging over multiple simulations to obtain better statistics, naturally, drives the cost of the Monte Carlo simulations up.)

While we note that we have not included effects such as local lattice distortions, the low computational cost of our calculations combined with the successful comparison of our results with existing literature suggests that our method is a highly efficient way of exploring the space of candidate multicomponent alloys, particularly when compared to high-throughput calculations on supercells using *ab initio* evaluations of energies, which use orders of magnitude more core hours.

## 4.7 Summary

Our results suggest that there is a set of simple underlying mechanisms driving atomic ASRO in these refractory HEAs. Pairs of $4d/5d$ elements which are isoelectronic interact weakly, mix well and make little or no contribution to ASRO in a material. Pairs of $4d/5d$ elements with a valence difference are favoured as nearest neighbours and drive B2-like ordering. For NbMoTa we find a B2-like (Nb, Ta; Mo) ordering, and for NbMoTaW we find a similar B2-like (Nb,Ta;Mo,W) ordering.

The addition of the $3d$ element V, with its smaller atomic size and narrower $d$-band drives a different, B32-like order, competing with and eventually dominating the B2-like ordering. We suggest that this small atom-big atom effect will be important in understanding phase stability in many $3d$-$4d$/$5d$ multicomponent alloys. Our calculations on VNbMoTaW also predict order emerging at a sufficiently high temperature that ASRO in this system may be experimentally observable given suitable heat treatment.

The agreement with existing literature is reassuring and we take it as evidence that our approach reliably describes the nature of order in multicomponent alloys at low computational cost, and can therefore be used to study other multicomponent systems to elucidate the nature of atomic arrangements.

## References


1. B. Cantor, I.T.H. Chang, P. Knight, A.J.B. Vincent, Mater. Sci. Eng.: A **375–377**, 213 (2004). https://doi.org/10.1016/j.msea.2003.10.257
2. J.W. Yeh, S.K. Chen, S.J. Lin, J.Y. Gan, T.S. Chin, T.T. Shun, C.H. Tsau, S.Y. Chang, Adv. Eng. Mater. **6**(5), 299 (2004). https://doi.org/10.1002/adem.200300567
3. Y. Zhang, T.T. Zuo, Z. Tang, M.C. Gao, K.A. Dahmen, P.K. Liaw, Z.P. Lu, Progress Mater. Sci. **61**, 1 (2014). https://doi.org/10.1016/j.pmatsci.2013.10.001
4. E.J. Pickering, N.G. Jones, Int. Mater. Rev. **61**(3), 183 (2016). https://doi.org/10.1080/09506608.2016.1180020
5. M.C. Gao, J.W. Yeh, P.K. Liaw, Y. Zhang (eds.), *High-Entropy Alloys* (Springer, Cham, 2016). https://doi.org/10.1007/978-3-319-27013-5
6. D.B. Miracle, O.N. Senkov, Acta Materialia **122**, 448 (2017). https://doi.org/10.1016/j.actamat.2016.08.081
7. E.P. George, D. Raabe, R.O. Ritchie, Nat. Rev. Mater. **4**(8), 515 (2019). https://doi.org/10.1038/s41578-019-0121-4
8. O. Senkov, G. Wilks, D. Miracle, C. Chuang, P. Liaw, Intermetallics **18**(9), 1758 (2010). https://doi.org/10.1016/j.intermet.2010.05.014
9. O. Senkov, G. Wilks, J. Scott, D. Miracle, Intermetallics **19**(5), 698 (2011). https://doi.org/10.1016/j.intermet.2011.01.004
10. A. Fernández-Caballero, J.S. Wróbel, P.M. Mummery, D. Nguyen-Manh, J. Phase Equilib. Diffus. **38**(4), 391 (2017). https://doi.org/10.1007/s11669-017-0582-3
11. Y. Zou, H. Ma, R. Spolenak, Nat. Commun. **6**(1), 7748 (2015). https://doi.org/10.1038/ncomms8748
12. Y. Zou, S. Maiti, W. Steurer, R. Spolenak, Acta Materialia **65**, 85 (2014). https://doi.org/10.1016/j.actamat.2013.11.049
13. X.G. Li, C. Chen, H. Zheng, Y. Zuo, S.P. Ong, npj Comput. Mater. **6**(1), 70 (2020). https://doi.org/10.1038/s41524-020-0339-0
14. M. Del Grosso, G. Bozzolo, H. Mosca, Physica B: Condens. Matter **407**(16), 3285 (2012). https://doi.org/10.1016/j.physb.2011.12.088
15. M.F. Del Grosso, G. Bozzolo, H.O. Mosca, J. Alloys Compd. **534**, 25 (2012). https://doi.org/10.1016/j.jallcom.2012.04.053
16. W.P. Huhn, M. Widom, JOM **65**(12), 1772 (2013). https://doi.org/10.1007/s11837-013-0772-3
17. M. Widom, W.P. Huhn, S. Maiti, W. Steurer, Metall. Mater. Trans. A **45**(1), 196 (2014). https://doi.org/10.1038/s41524-019-0195-y
18. F. Körmann, M. Sluiter, Entropy **18**(8), 403 (2016). https://doi.org/10.3390/e18080403
19. F. Körmann, A.V. Ruban, M.H. Sluiter, Mater. Res. Lett. **5**(1), 35 (2017). https://doi.org/10.1080/21663831.2016.1198837
20. P. Singh, A.V. Smirnov, D.D. Johnson, Phys. Rev. Mater. **2**(5), 055004 (2018). https://doi.org/10.1103/PhysRevMaterials.2.055004

21. A.D. Kim, M. Widom, Phys. Rev. Mater. **7**(6), 063803 (2023). https://doi.org/10.1103/PhysRevMaterials.7.063803
22. A. Fantin, A.M. Manzoni, H. Springer, R.D. Kamachali, R. Maaß, Mater. Res. Lett. **12**(5), 346 (2024). https://doi.org/10.1080/21663831.2024.2326014
23. J.S. Faulkner, G.M. Stocks, Phys. Rev. B **21**(8), 3222 (1980). https://doi.org/10.1103/PhysRevB.21.3222
24. J.S. Faulkner, G.M. Stocks, Y. Wang, *Multiple Scattering Theory: Electronic Structure of Solids*, 1st edn. (IOP Publishing, Bristol, UK, 2018). https://doi.org/10.1088/2053-2563/aae7d8
25. D.D. Johnson, D.M. Nicholson, F.J. Pinski, B.L. Györffy, G.M. Stocks, Phys. Rev. B **41**(14), 9701 (1990). https://doi.org/10.1103/PhysRevB.41.9701
26. M. Hoffmann, A. Ernst, W. Hergert, V.N. Antonov, W.A. Adeagbo, R.M. Geilhufe, H. Ben Hamed, Phys. Status Solidi (b) **257**(7), 1900671 (2020). https://doi.org/10.1002/pssb.201900671
27. G.M. Stocks, W.M. Temmerman, B.L. Gyorffy, Phys. Rev. Lett. **41**(5), 339 (1978). https://doi.org/10.1103/PhysRevLett.41.339
28. J.P. Perdew, Y. Wang, Phys. Rev. B **45**(23), 13244 (1992). https://doi.org/10.1103/PhysRevB.45.13244
29. C.D. Woodgate, J.B. Staunton, Phys. Rev. Mater. **7**(1), 013801 (2023). https://doi.org/10.1103/PhysRevMaterials.7.013801
30. L. Vegard, Zeitschrift für Physik **5**(1), 17 (1921). https://doi.org/10.1007/BF01349680
31. A.R. Denton, N.W. Ashcroft, Phys. Rev. A **43**(6), 3161 (1991). https://doi.org/10.1103/PhysRevA.43.3161
32. F.J. Pinski, B. Ginatempo, D.D. Johnson, J.B. Staunton, G.M. Stocks, B.L. Gyorffy, Phys. Rev. Lett. **66**(6), 766 (1991). https://doi.org/10.1103/PhysRevLett.66.766
33. W. Hume-Rothery, *Atomic Theory for Students of Metallurgy* (The Institute of Metals, London, 1969)
34. M. Muzyk, D. Nguyen-Manh, K.J. Kurzyd lowski, N.L. Baluc, S.L. Dudarev, Phys. Rev. B **84**(10), 104115 (2011). https://doi.org/10.1103/PhysRevB.84.104115
35. T. Kostiuchenko, F. Körmann, J. Neugebauer, A. Shapeev, npj Comput. Mater. **5**(1), 55 (2019). https://doi.org/10.1038/s41524-019-0195-y
36. L.F. Zhu, P. Srinivasan, Y. Gong, T. Hickel, B. Grabowski, F. Körmann, J. Neugebauer, Phys. Rev. B **109**(9), 094110 (2024). https://doi.org/10.1103/PhysRevB.109.094110
37. X. Liu, J. Zhang, J. Yin, S. Bi, M. Eisenbach, Y. Wang, Comput. Mater. Sci. **187**, 110135 (2021). https://doi.org/10.1016/j.commatsci.2020.110135
38. A. Stukowski, Model. Simul. Mater. Sci. Eng. **18**(1), 015012 (2010). https://doi.org/10.1088/0965-0393/18/1/015012
39. S.A. Kube, S. Sohn, D. Uhl, A. Datye, A. Mehta, J. Schroers, Acta Materialia **166**, 677 (2019). https://doi.org/10.1016/j.actamat.2019.01.023
40. L. Lannelongue, J. Grealey, M. Inouye, Adv. Sci. **8**(12), 2100707 (2021). https://doi.org/10.1002/advs.202100707

# Chapter 5
# Complex, Multiphase Behaviour in the $Ti_x$NbMoTaW and $Ti_x$VNbMoTaW Refractory High-Entropy Alloys

**Abstract** We study the influence of Ti additions on the predicted atomic arrangements and microstructure of the NbMoTaW and VNbMoTaW refractory high-entropy alloys. The addition of Ti is understood experimentally to improve the room-temperature ductility of these alloys, but also to lead to the emergence of chemical inhomogeneities during processing. Our modelling suggests that the addition of Ti leads to competition between phase segregation and phase ordering in these systems, and also to increased ordering temperatures when compared to the systems with no Ti present. Lattice-based Monte Carlo simulations evidence competing interactions and complex, multiphase behaviour in the Ti-containing systems. These results give fundamental physical insight into the experimentally observed microstructure of these systems, and serve as further validation of our modelling approach.

## 5.1 Introduction

In the previous chapter, we examined the phase stability of the prototypical refractory high-entropy alloys, VNbMoTaW, NbMoTaW, and selected subsystems [1, 2]. These well-studied systems served as an important benchmark, allowing the proposed methodology to be compared to existing literature. However, despite their excellent elevated temperature performance, in practice these materials are highly brittle at room temperature, which limits their potential applicability [3, 4]. One suggested route to improve room-temperature ductility in these materials is by alloying with Ti, forming the systems $Ti_x$NbMoTaW and $Ti_x$VNbMoTaW [3]. However, experimental studies of these compositions have shown that the addition of Ti introduces microstructural inhomogeneities, suggesting competing interactions and complex, multiphase behaviour.

In this chapter, we explore this aspect systematically within the our proposed modelling framework. We study the phase behaviour of the $Ti_x$NbMoTaW and $Ti_x$VNbMoTaW systems via first-principles calculations, our linear-response analysis, and lattice-based Monte Carlo simulations. We are able to demonstrate that the addition of Ti leads to competition between phase segregation and phase ordering in these systems. Further, our Monte Carlo simulations show complex, multiphase



C. D. Woodgate, *Modelling Atomic Arrangements in Multicomponent Alloys*,
Springer Series in Materials Science 346,
https://doi.org/10.1007/978-3-031-62021-8_5

behaviour emerging in these systems, which correlates well with experimental observations. These results therefore give theoretical understanding of the experimentally observed phase behaviour of these systems.

The rest of this chapter is structured in a very similar manner to Chap. 4. First, in Sect. 5.2, we discuss the nature of the underlying crystal lattice with increasing Ti concentration (hcp *versus* bcc), before examining the electronic structure of the disordered solid solution as modelled within the CPA. Then, in Sect. 5.3, we perform our perturbative analysis, examining how the CPA reference medium responds to an applied chemical fluctuation. In Sect. 5.4, we map the results of this linear-response analysis to an effective pairwise model describing the internal energy of the alloy. Then, in Sect. 5.5, we perform lattice-based Monte Carlo simulations for the considered systems using this pairwise model. Finally, in Sect. 5.6, we summarise the results and draw conclusions.

## 5.2 Electronic Structure

As for the previous chapter, we begin by performing a self-consistent DFT calculation to model the electronic structure of the disordered solid solution. Again, we use the all-electron *Hutsepot* code [6] to construct the self-consistent potentials of the KKR formulation of DFT, using the CPA to average over chemical disorder. We perform scalar-relativistic calculations within the atomic sphere approximation (ASA) [7], employing an angular momentum cutoff of $l_{\text{max}} = 3$ for basis set expansions, a $20 \times 20 \times 20$ **k**-point mesh for integrals over the Brillouin zone, and a 24 point semi-circular Gauss-Legendre grid in the complex plane to integrate over valence energies. We use the local density approximation and the exchange-correlation functional is that of Perdew-Wang [8]. Rather than optimise lattice parameters, in this chapter we use experimentally-determined lattice constants to illustrate the robustness of the approach. Therefore, bcc lattice parameters for NbMoTaW and VNbMoTaW are set at 3.226 Å and 3.183 Å respectively, consistent with their experimental values [1, 4]. For TiNbMoTaW and TiVNbMoTaW (*i.e.*, the case $x = 1$) these values are set at 3.240 Å and 3.209 Å respectively, again consistent with experimentally determined values [3]. For intermediate values of $x$ we interpolate linearly between these using Vegard's law [9].

Previous CALPHAD modelling has suggested the possibility of the emergence of an hcp phase at low temperatures with increasing Ti concentration [4], in alignment with Ti typically forming on an hcp lattice at room temperature. To investigate this aspect, we perform a transformation from the bcc structure to the hcp structure conserving the overall volume-per-atom and compute the difference in total energy per atom between the bcc and hcp structures as a function of $x$, the results of which are visualised in Fig. 5.1. Within the ASA, using the CPA to average over compositional disorder, we find that the bcc structure is consistently favoured over the hcp structure in the region $0 \leq x \leq 2$, *i.e.*, up to and beyond the range of $x$ considered in this

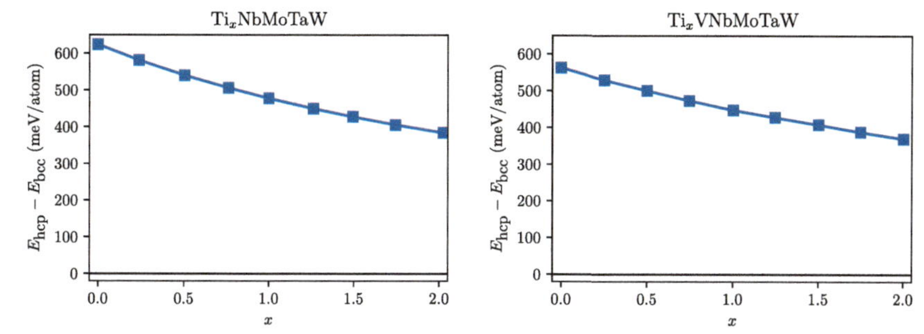


**Fig. 5.1** Comparison of the difference in total energy-per-atom as a function of $x$ for $Ti_x$NbMoTaW (left) and $Ti_x$VNbMoTaW (right) comparing the bcc and hcp structures, computed using the all-electron KKR-CPA *Hutsepot* code. Across the range of $x$ considered, bcc is clearly the favoured structure for both compositions

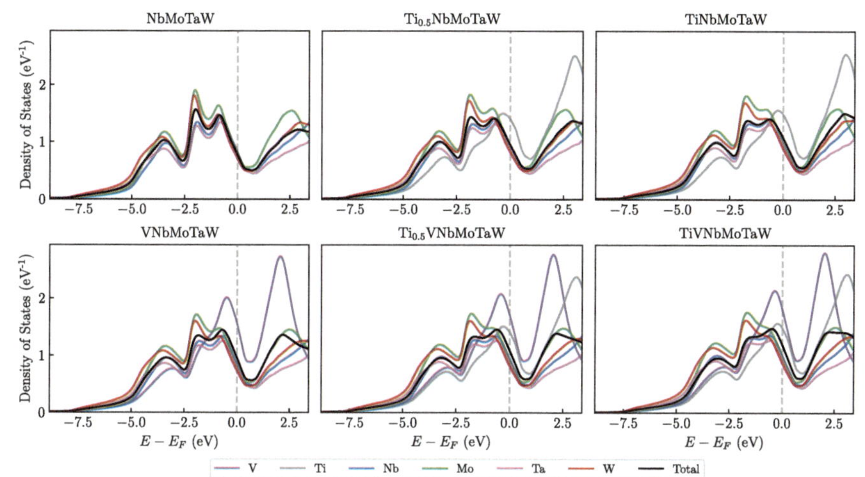


**Fig. 5.2** Comparison of the total and species-resolved density of states for different values of $x$ for the disordered solid solution modelled within the coherent potential approximation for both $Ti_x$NbMoTaW and $Ti_x$VNbMoTaW. The Fermi level, $E_F$, is indicated by a vertical, grey, dashed line. The 3*d* elements V and Ti have species-resolved curves which differ substantially in character from the 4*d*/5*d* elements Nb, Mo, Ta, and W, suggesting the potential for stronger ordering tendencies

chapter. We therefore only consider the alloy on a bcc lattice for the remainder of this chapter.

Proceeding, in Fig. 5.2 we visualise the total and species-resolved density of states (DoS) for both systems within the CPA for three indicative values of $x$ when the underlying lattice is bcc. The total DoS is given by the weighted average of the species-resolved curves. As for the refractory systems of Chap. 4, it can be seen that pairs of chemical species which are isoelectronic, *e.g.*, Nb/Ta and Mo/W, have species-resolved curves which lie almost on top of one another, which we expect to lead to weak correlations between these pairs of elements. Compared to the $4d$ and $5d$ elements, the two $3d$ elements considered here, V and Ti, can be seen to have narrower $d$ bands with significantly different profiles around $E_F$, which we expect to lead to strong ordering tendencies between these and other elements.

Another important feature to discuss is the charge-transfer between elements, as noted in Chap. 4. In an alloy system where there is an atomic size discrepancy, where the lattice parameter sits somewhere between that of the pure elements (approximately in accordance with Vegard's law), there is often a transfer of charge from the large atoms to the smaller ones, corresponding to the charge density associated with the 'large' atom spilling over into the Winger-Seitz cell associated with the 'small' atom. In our calculations, we find that this effect is generally strongest for V and Ta, with V gaining charge and Ta losing it. For example, in the five-component VNbMoTaW, V (proton number 23) has an associated average charge per atom of 23.177 $e$, while Ta (proton number 73) has an associated charge per-atom of 72.901 $e$. This is consistent with V being the 'smallest' atom and Ta one of the 'largest' for the systems considered here.

## 5.3 Linear Response

Proceeding, we perform a linear response calculation to assess the dominant atomic short- and long-range order in these systems, as outlined in Chaps. 2 and 3. Visualised in Fig. 5.3a is the raw $S^{(2)}_{\alpha\alpha'}(\mathbf{k})$ for the considered systems. Then, in Fig. 5.3b, we use this data to construct the chemical stability matrix and visualise its eigenvalues around the irreducible Brillouin zone (IBZ) of the bcc lattice a temperature of $T = 1000$ K. Note that, as before, for an $s$-species alloy, there are $s - 1$ eigenvalues due to the constraint that the overall concentrations of each species be conserved.

When the raw $S^{(2)}_{\alpha\alpha'}(\mathbf{k})$ is considered (Fig. 5.3a) we see that, by far the strongest atom-atom correlations are between the $3d$ elements (Ti, V) and the $4d/5d$ elements (Nb, Mo, Ta, W). This is in alignment with the results of the previous chapter, and related to the relative atomic size differences, bandwidth differences, and associated charge transfer between $3d$ and $4d/5d$ elements. In addition, we stress that the valence difference Ti, a group 4 element, and the five refractory elements, which are groups 5 and 6, will also play a role in strengthening atomic ordering tendencies.

We then move on to consider the eigenvalues of the chemical stability matrix, Fig. 5.3b. As a reminder, to interpret these concentration wave modes, there are three key features to consider. The first is the shape of the modes, as these tell us about the nature and strength of atom-atom interactions in the disordered phase. Strongly varying modes are associated with strong interactions, while weakly varying modes are associated with chemical species interacting weakly. The second feature is the location of the minima, as these tell us about the dominant correlations in the disordered phase, and to infer the likely chemical ordering. Finally, it is necessary to know the chemical polarisation of a mode, *i.e.*, $\Delta c_\alpha$ to understand the chemical species to which it relates. Some sample eigenvalues, and chemical polarisations

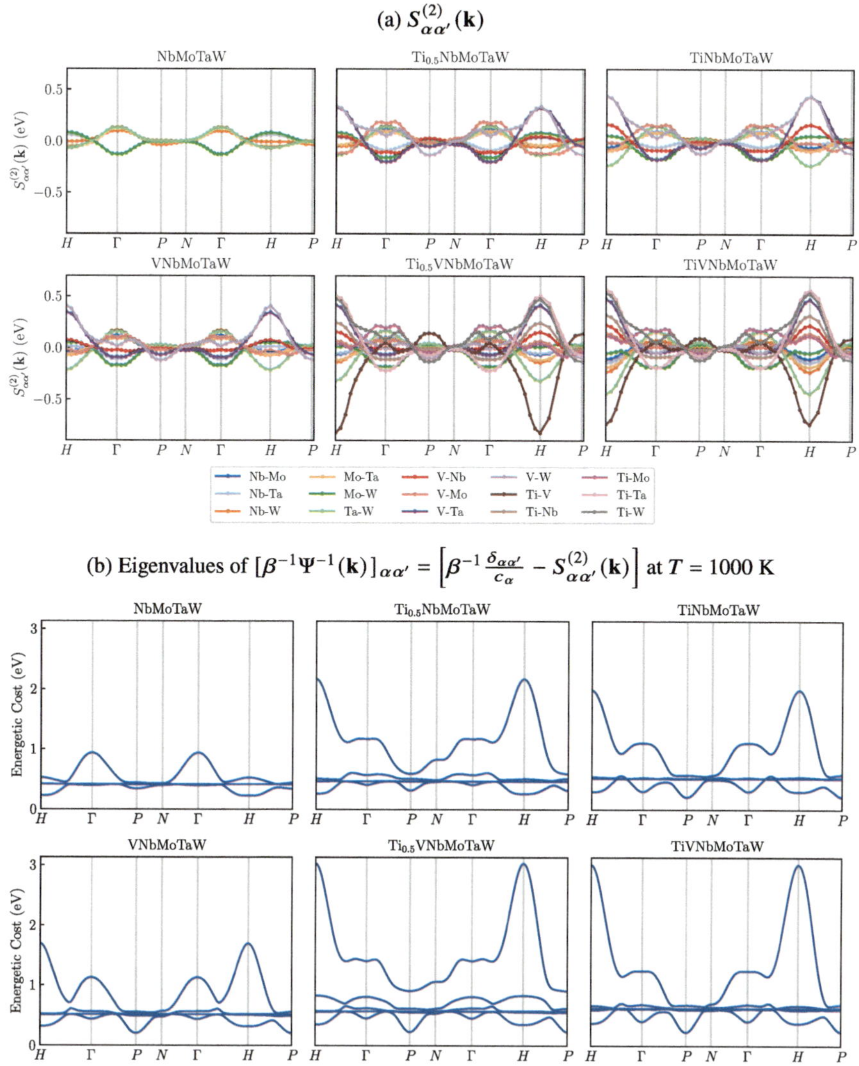


**Fig. 5.3** Plots of the raw $S^{(2)}_{\alpha\alpha'}(\mathbf{k})$ data (top) and eigenvalues of the chemical stability matrix (bottom), around the irreducible Brillouin Zone of the body-centred cubic lattice. The chemical stability matrix is evaluated at $T = 1000$ K. The $S^{(2)}_{\alpha\alpha'}(\mathbf{k})$ includes the Onsager correction. Without the addition of Ti, interactions are comparatively weak, and the minimum eigenvalues lie at $H$ and $P$, indicative of B2 and B32 ordering, respectively. However, with increasing Ti concentration, for both systems, stronger correlations emerge, as does an additional minimum eigenvalue $\Gamma$, indicative of the potential for phase segregation

at a few high-symmetry points of the IBZ for TiNbMoTaW and TiVNbMoTaW are provided in Table 5.1, while the remaining data can be found in the repository associated with this book.

Considering first the $Ti_x$NbMoTaW system, the top row of Fig. 5.3b, we see that for the case $x = 0$, *i.e.*, NbMoTaW, we have three clear modes emerging. The first is a mode dipping at $H$ and peaking at $\Gamma$, which is associated with B2 ordering tendencies and, in concentration space, is polarised to suggest that Nb and Ta will sit on one sublattice, while Mo and W will sit on the other. The two other modes present are near-flat, and are associated with the isoelectronic pairs Nb-Ta and Mo-W, which are very weakly correlated in our calculations. On adding Ti by increasing $x$, however, we see the emergence of a new, strongly varying mode, introducing competing minima at $P$ and $\Gamma$, associated with B32 ordering and phase segregation respectively. Notably, the chemical polarisation of the concentration wave mode at the $\Gamma$ point, as shown in Table 5.1, suggests Ti, and Mo segregating from Nb, Ta, and W, which is entirely consistent with the experimentally observed segregation in [4]. These results therefore confirm that the addition of Ti produces competing interactions in this system, with the potential for a variety of chemical orderings and/or phase segregation.

Proceeding, we move on to the results for the $Ti_x$VNbMoTaW alloy, *i.e.*, the bottom row of Fig. 5.3b. For the case $x = 0$, there is an additional mode compared to NbMoTaW, which dips at $P$ and peaks at $H$, and is associated with a B32 ordering between V and the other elements present. This was analysed in detail in Chap. 4 and [10] and is associated with the large charge-transfer between V and the other elements present. Again, upon the addition of Ti, however, a new, strongly varying mode materialises and introduces a competing minimum at the $\Gamma$ point, indicative of phase segregation. As shown in Table 5.1, this minimum has a chemical polarisation suggesting Ti, V, and Mo segregating away from Nb, Ta, and W, again consistent with the experimental data [3, 11]. As before, we interpret these data as suggesting

**Table 5.1** Comparison of eigenvalues and their associated chemical polarisations for $Ti_x$NbMoTaW and $Ti_x$VNbMoTaW for $x = 1$. All are evaluated at $T = 1000$ K. For both systems, for the special points $H$, $\Gamma$, and $P$ (indicative of B2 ordered, phase separation, and B32 ordering, respectively), the eigenvalues are close in value, indicative of competition between different concentration wave modes. At the $\Gamma$ point, the chemical polarisation is dominated by Ti, suggesting a tendency for Ti to segregate away from other elements

| System | **k**-point | Eigenvalue (eV) | $\Delta c_{\text{Ti}}$ | $\Delta c_{\text{V}}$ | $\Delta c_{\text{Nb}}$ | $\Delta c_{\text{Mo}}$ | $\Delta c_{\text{Ta}}$ | $\Delta c_{\text{W}}$ |
|---|---|---|---|---|---|---|---|---|
| TiNbMoTaW | $H$ | 0.290 | −0.107 | | −0.361 | 0.636 | −0.553 | 0.386 |
| | $\Gamma$ | 0.297 | −0.784 | | 0.168 | −0.163 | 0.508 | 0.271 |
| | $P$ | 0.227 | −0.828 | | −0.011 | 0.101 | 0.246 | 0.493 |
| TiVNbMoTaW | $H$ | 0.368 | −0.252 | 0.167 | −0.361 | 0.589 | −0.530 | 0.388 |
| | $\Gamma$ | 0.402 | −0.611 | −0.331 | 0.138 | −0.144 | 0.589 | 0.360 |
| | $P$ | 0.253 | −0.605 | −0.487 | 0.099 | 0.172 | 0.314 | 0.508 |

that Ti does not mix well with the other elements present and introduces competing chemical interactions in the system.

For each of the considered systems, via application of the Landau theory outlined in Chap. 2, we estimate a transition temperature by computing the temperature for which the lowest-lying eigenvalue passes through zero. We also provide its chemical polarisation and associated wavevector, as these elucidate the nature of the ordered phase. These results are tabulated in Table 5.2. In summary for both systems, increasing Ti concentration leads to higher predicted ordering temperatures, and alters the chemical polarisation and/or wavevector describing the chemical ordering. It is evident that Ti drives strong atom-atom correlations in these systems. We note that, although our predicted ordering is B32, the above analysis emphasises that this ordering competes closely with phase segregation when Ti is present in the competition. This suggests that further atomistic modelling is required to robustly assess the phase stability of these systems.

We emphasise that both the temperature and nature of our predicted B2 ordering in NbMoTaW and predicted B32 ordering in VNbMoTaW are in good agreement with other DFT-based computational studies [12–15], including those using GGA functionals, and we take this as evidence that the LDA adequately captures the relevant physics in these systems. For the 5-component $Ti_x$NbMoTaW and 6-component $Ti_x$VNbMoTaW systems, there is limited DFT data with which to compare, but we take the alignment of our results with the experimental data as reassurance that our description of the electronic structure of these systems is adequate.

### 5.3.1 The Effect of Lattice Expansion

In the previous chapter, we simulated the NbMoTaW and VNbMoTaW systems using LDA-optimised lattice parameters, which are reduced compared to their experimental values. However, in this chapter, we instead opt to use the experimental values

**Table 5.2** Predicted ordering temperatures ($T_{us}$), wavevectors ($\mathbf{k}_{us}$), and chemical polarisations for the $Ti_x$NbMoTaW and $Ti_x$VNbMoTaW systems for $x = 0, 0.5, 1$. The addition of Ti generally increases predicted ordering temperatures on account of strong correlations between Ti and other elements

| System | $x$ | $T_{us}$ (K) | $\mathbf{k}_{us}$ ($\frac{2\pi}{a}$) | $\Delta c_{Ti}$ | $\Delta c_V$ | $\Delta c_{Nb}$ | $\Delta c_{Mo}$ | $\Delta c_{Ta}$ | $\Delta c_W$ |
|---|---|---|---|---|---|---|---|---|---|
| $Ti_x$NbMoTaW | 0 | 557 | (0, 0, 1) | | | −0.421 | 0.566 | −0.568 | 0.424 |
| | 0.5 | 584 | $(\frac{1}{2}, \frac{1}{2}, \frac{1}{2})$ | −0.725 | | −0.168 | 0.047 | 0.217 | 0.629 |
| | 1 | 673 | $(\frac{1}{2}, \frac{1}{2}, \frac{1}{2})$ | −0.828 | | −0.011 | 0.101 | 0.246 | 0.493 |
| $Ti_x$VNbMoTaW | 0 | 698 | $(\frac{1}{2}, \frac{1}{2}, \frac{1}{2})$ | | −0.828 | −0.005 | 0.090 | 0.247 | 0.495 |
| | 0.5 | 726 | $(\frac{1}{2}, \frac{1}{2}, \frac{1}{2})$ | −0.397 | −0.656 | 0.054 | 0.159 | 0.298 | 0.542 |
| | 1 | 709 | $(\frac{1}{2}, \frac{1}{2}, \frac{1}{2})$ | −0.605 | −0.487 | 0.099 | 0.172 | 0.324 | 0.507 |

**Table 5.3** Effect of lattice expansion/contraction on predicted atomic ordering temperatures for NbMoTaW and VNbMoTaW. It can be seen that, around the experimental value of the lattice constant, small changes to the lattice spacing do not significantly alter the predictions of the linear reponse analysis

| Composition | $a$ (Å) | $a/a_{\text{exp.}}$ | $T_{\text{ord}}$ | $\mathbf{k}_{\text{ord}}$ $(2\pi/a)$ | $\Delta c_{\text{V}}$ | $\Delta c_{\text{Nb}}$ | $\Delta c_{\text{Mo}}$ | $\Delta c_{\text{Ta}}$ | $\Delta c_{\text{W}}$ |
|---|---|---|---|---|---|---|---|---|---|
| NbMoTaW | 3.161 | 0.98 | 560 | $(0, 0, 1)$ | | −0.389 | 0.590 | −0.591 | 0.389 |
| | 3.226 | 1.00 | 557 | $(0, 0, 1)$ | | −0.421 | 0.566 | −0.568 | 0.424 |
| | 3.290 | 1.02 | 548 | $(0, 0, 1)$ | | −0.450 | 0.540 | −0.546 | 0.456 |
| VNbMoTaW | 3.119 | 0.98 | 740 | $(\frac{1}{2}, \frac{1}{2}, \frac{1}{2})$ | −0.837 | 0.010 | 0.115 | 0.224 | 0.486 |
| | 3.183 | 1.00 | 698 | $(\frac{1}{2}, \frac{1}{2}, \frac{1}{2})$ | −0.828 | −0.005 | 0.090 | 0.247 | 0.495 |
| | 3.247 | 1.02 | 627 | $(\frac{1}{2}, \frac{1}{2}, \frac{1}{2})$ | −0.837 | 0.014 | 0.113 | 0.225 | 0.486 |

directly. It is not unreasonable to ask how this change in lattice spacing affects our modelling results.

To address this question, we consider the NbMoTaW and VNbMoTaW systems. Tabulated in Table 5.3 are predicted ordering temperatures, wavevectors, and concentration wave modes from the linear-response theory obtained for these two systems simulated across a range of lattice spacings. For both systems, it is found that a decrease in lattice spacing results in an increased atomic ordering temperatures, while an increased lattice spacing results in decreased atomic ordering temperatures. The effect is larger for VNbMoTaW than for NbMoTaW, which we attribute to the larger charge transfers between elements in the V-containing system. The charge transfers and associated screened Coulombic interaction between elements will be heavily influenced by choice of lattice spacing, being weakened with increased lattice spacing, while the band mechanisms driving ordering in NbMoTaW appear to be less sensitive to small deviations from the experimental values of lattice spacings. However, in both cases, the predicted orderings (B2 for NbMoTaW, B32 for VNbMoTaW) remain consistent, and the chemical polarisations remain very numerically similar. This serves as reassurance that our results are robust, although it may be that taking into account thermal expansion of the lattice could reduce predicted atomic ordering temperatures.

## 5.4 Pairwise Interactions

From the linear response calculations performed in reciprocal space, we are able to backwards Fourier transform and recover a pairwise, real-space interaction (effective pair interaction) as described in Chap. 2. In Fig. 5.4a, b, we visualise the recovered $V^{(n)}_{\alpha\alpha'}$ as a function of radial distance for the $Ti_x$NbMoTaW and $Ti_x$VNbMoTaW alloys respectively, for a range of values of $x$. As in the previous chapter, it can be seen that interactions are dominated by the first two coordination shells, and tail off quickly

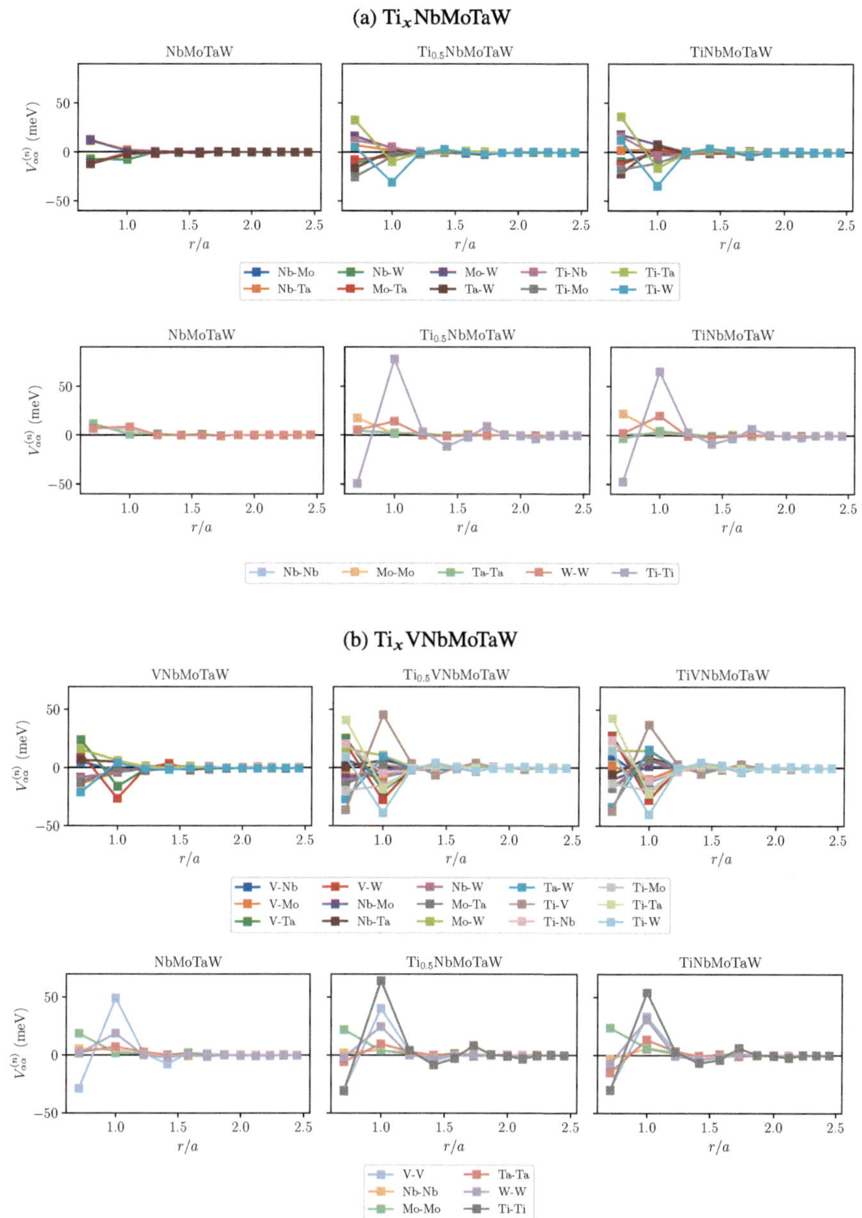


**Fig. 5.4** Visualisation of $V_{\alpha\alpha'}^{(n)}$ for the $Ti_x$VNbMoTaW refractory high-entropy alloy for selected values of $x$, fitted to the first twelve coordination shells. It can be seen that interactions are dominated by the first two coordination shells, and are negligible beyond sixth-nearest neighbour distance

with increasing distance. Indeed, a fit to the first four coordination shells captures the reciprocal space data with good accuracy, and no significant improvement to the fit is made once interactions up to and including the sixth coordination shell are considered. For the purposes of the subsequent atomistic modelling, we therefore truncate our fit at the sixth coordination shell. The explicit values of $V^{(n)}_{\alpha\alpha'}$ are then tabulated in Appendix C, Tables C.4, C.5, C.6. We note that a fit to the first six coordination shells is a small difference from [5], which only included interactions up to and including the fourth coordination shell. As we shall see shortly, these weak interactions on the fifth and sixth coordination shell do not significantly alter the predicted phase behaviour.

For both the $Ti_x$NbMoTaW and $Ti_x$VNbMoTaW systems, we find that, typically, the strongest interactions are between Ti, V, and the other elements present, while $4d/5d$ pairs such as Nb-Ta and Mo-W, which are isoelectronic, interact weakly. Some interactions are stronger on the second coordination shell than the first, suggestive of B32-like ordering tendencies, consistent with the linear-response analysis. In addition, we note that our fitted interactions exhibit clear dependence on system composition, which reflects the importance of a full treatment of the multicomponent system rather than extrapolating interactions from binary subsystems. This composition-dependence is understood to originate in the changing nature of the electronic structure of these materials as Ti is added in increasing concentrations.

## 5.5 Atomistic Modelling

Proceeding, we perform lattice-based Monte Carlo simulations of both the $Ti_x$NbMoTaW and $Ti_x$VNbMoTaW systems for $x = 0$, 0.5, 1, using simulated annealing, as outlined in Chap. 2. As for the previous chapter, all simulation cells consisted of a system of $16 \times 16 \times 16$ cubic bcc unit cells for a total of 8192 atoms in the simulation cell. At each temperature, following an initial burn-in period to achieve equilibrium, statistics were gathered over a run of $10^4$ Monte Carlo steps per atom. All data were averaged across an ensemble of 16 simulations, which enables an extraction of an uncertainty on key quantities such as the specific heat capacity of the simulation.

Visualised in Figs. 5.5 and 5.6 are plots of the Warren-Cowley ASRO parameters at first- and second-nearest-neighbour distance and specific heat capacity as a function of temperature for $Ti_x$NbMoTaW and $Ti_x$VNbMoTaW respectively. The top row of panels for each subfigure shows Warren-Cowley parameters excluding Ti, while the bottom row shows the Warren-Cowley parameters for correlations between Ti and the other elements present.

Considering first the results for $Ti_x$NbMoTaW (Fig. 5.5), we see that there is a clear trend of increasing transition temperature with increasing Ti concentration, as the initial peak in SHC is moved to higher temperatures. In the NbMoTaW system, as discussed in our earlier work [10], the initial peak in the SHC is associated with a B2 ordering, with Nb and Ta atoms preferentially sitting on one sublattice, and

(a) Nearest-neighbour distance.

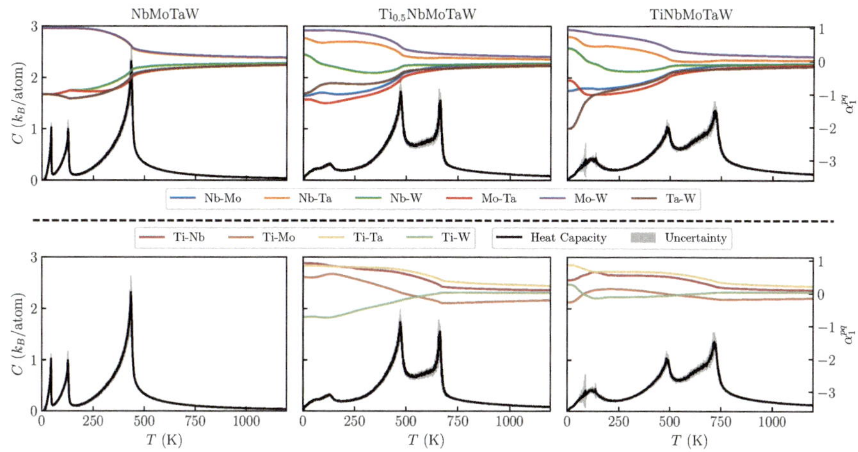


(b) Second-nearest-neighbour distance.

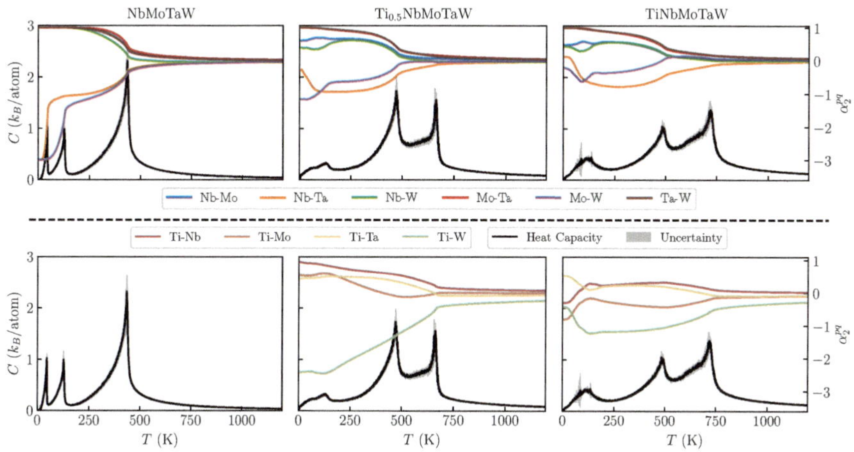


**Fig. 5.5** Plots of the Warren-Cowley ASRO parameters, $\alpha_n^{pq}$, and specific heat capacity, $C$, for $Ti_x$NbMoTaW as a function of temperature, calculated from an ensemble of 16 lattice-based Monte Carlo simulations, each containing 8192 atoms. Figure 5.5a shows the Warren-Cowley parameters on the first coordination shell (*i.e.*, at nearest-neighbour distance) while 5.5b shows them on the second coordination shell (*i.e.*, at second-nearest-neighbour distance). The top row of each subfigure shows atom-atom correlations excluding Ti, while the bottom row shows correlations between Ti and other elements. The addition of Ti can be seen to introduce an additional Ti-driven ordering compared to the NbMoTaW system

Mo and W atoms preferentially sitting on the other. This predicted B2 ordering is in good agreement with other computational studies of this system [12, 14, 15, 17]. The peak in SHC associated with the B2 ordering is maintained as Ti is added to

(a) Nearest-neighbour distance.

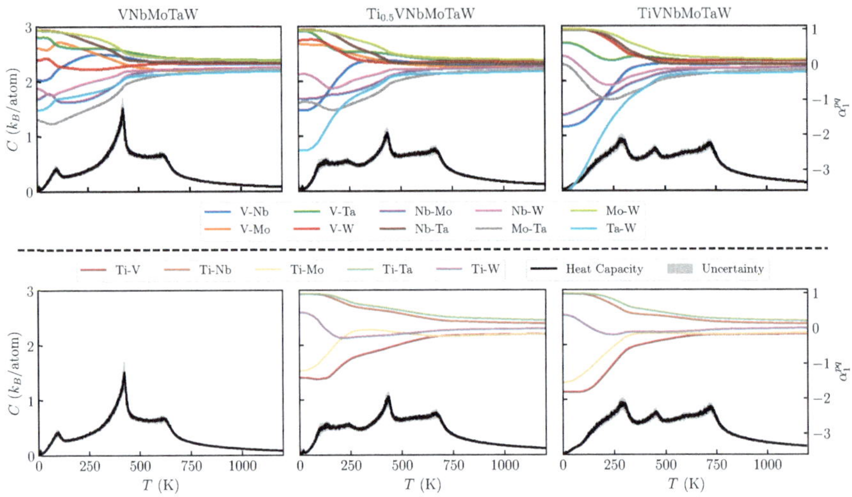


(b) Second-nearest-neighbour distance.

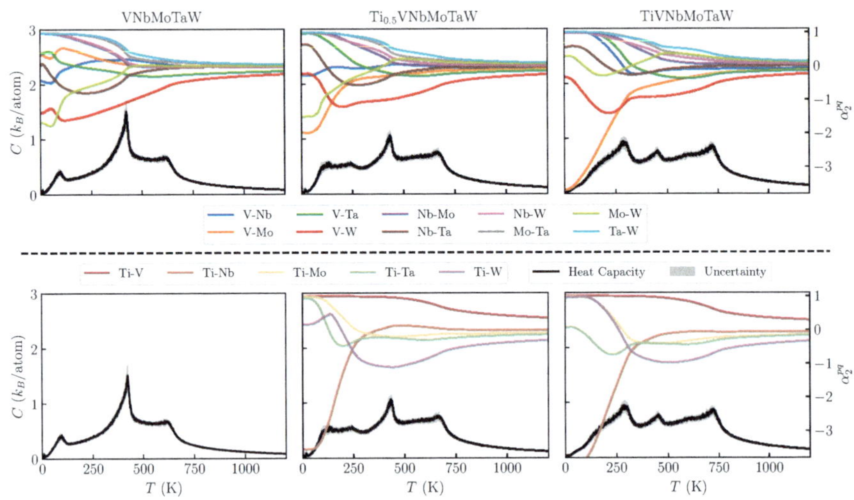


**Fig. 5.6** Plots of the Warren-Cowley ASRO parameters, $\alpha_n^{pq}$, and specific heat capacity, $C$, for $Ti_x$VNbMoTaW as a function of temperature extracted from our Monte Carlo simulations. Figure 5.6a shows the Warren-Cowley parameters on the first coordination shell while 5.6b shows them on the second coordination shell. The top row of each subfigure shows atom-atom correlations excluding Ti, while the bottom row shows correlations between Ti and other elements. The addition of Ti can be seen to introduce an additional Ti-driven ordering compared to the VNbMoTaW system

the system, but a new, additional peak at higher temperatures emerges, associated with correlations between Ti and the other elements present. In particular, in our simulations, at nearest-neighbour distance, Ti favours pairing with W, and avoids Mo, Ta, and Nb. At second-nearest-neighbour distance, Ti again favours pairing with W, while Mo-W and Nb-Ta pairs are also favoured. These results are consistent with the earlier linear response analysis suggesting that the introduction of Ti produces competing interactions and eventual phase segregation in this system.

Moving on to the results for the senary $Ti_x$VNbMoTaW system (Fig. 5.6), we see a similar picture emerging. Without Ti present, there is known to be emergent B32-like ordering in this system [10, 13], which is detected in both our linear response calculation and Monte Carlo simulations, indicated here by the peak in SHC between 500 and 750 K. As for the $Ti_x$NbMoTaW system, however, the introduction of Ti produces an additional peak in the SHC curves at higher temperature, which is associated with changing between Ti and the other elements present. Our modelling suggests that, at nearest-neighbour distance, Ti-V pairs are favoured, while Ti-Ta pairs are disfavoured. We also find that Ta-W and Mo-Ta nearest-neighbour pairs (valence difference) are favoured, while Mo-W and Nb-Ta nearest-neighbour pairs (isoelectronic) are disfavoured, reminiscent of the B2 ordering for NbMoTaW. At second-nearest-neighbour distance, V-W and Ti-W pairs are favoured, suggestive of B32-like ordering tendencies, while Ti-V, Ta-W, and Mo-W pairs are disfavoured.

These systems have highly complex potential energy surfaces and subtle phase behaviour. Therefore, to better interpret the above analysis, in Figs. 5.7 and 5.8, we visualise sample equilibrated configurations at selected temperatures for both considered systems for $x = 0$ and $x = 1$. These give intuitive insight into the complexity of Figs. 5.5 and 5.6; apart from NbMoTaW, there is no clear single-phase low temperature arrangement of atoms favoured. Instead, with decreasing temperature, we see emergent multiphase behaviour, with patches of binary/ternary ordered structures condensing out of the disordered phase. These results, therefore emphasise that the addition of Ti not only drives up predicted transition temperatures, but also that this alloying addition leads to competition between phase segregation and phase ordering. These results are therefore consistent with our earlier perturbative analysis. In addition, the observed multiphase behaviour is in alignment with experimentally-observed phase segregation in these systems.

For both systems, there are some small quantitative differences between the results of [5] and those presented here, on account of the fact that the modelling presented in this chapter includes interactions fitted to the first six coordination shells, while [5] used interactions fitted to the first four coordination shells. The weak interactions on the fifth and sixth coordination shell do not substantially qualitatively alter the nature of our results, however, and we take this as reassurance that these weaker interactions on further coordination shells are not particularly important for the systems considered here.

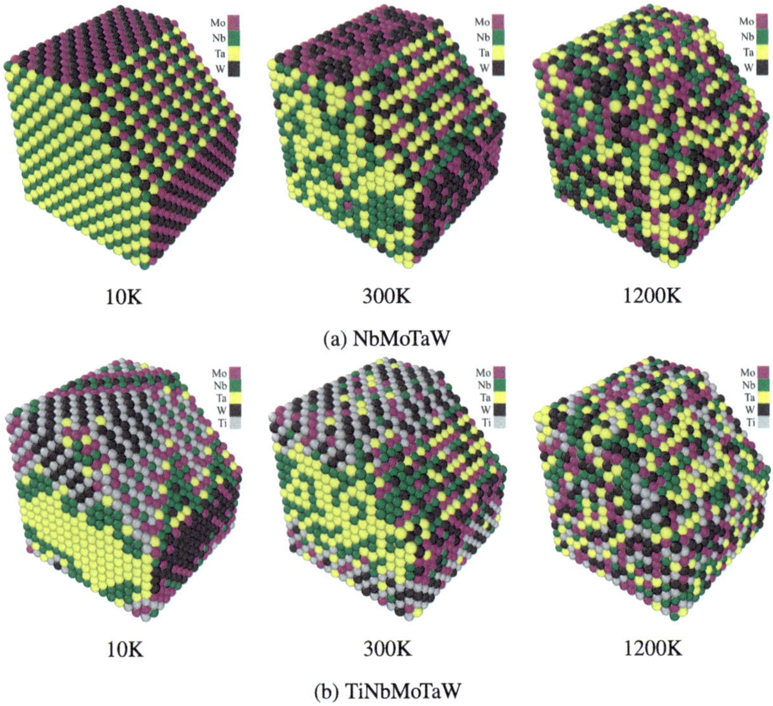


**Fig. 5.7** Visualisation of indicative equilibrium configurations for NbMoTaW (*i.e.*, $Ti_x$NbMoTaW, $x = 0$) and TiNbMoTaW (*i.e.*, $Ti_x$NbMoTaW, $x = 0$) at a selection of temperatures. Ti, Nb, Mo, Ta, and W are coloured grey, green, pink, yellow, and black, respectively. A cut has been made through the simulation cell to make the configurations easier to interpret. While the NbMoTaW system can be seen to exhibit a clear B2 ordering leading to an eventual Heusler-like ground state, the addition of Ti leads to more complex, multiphase behaviour. Images generated using Ovito [16]

## 5.6 Summary

In summary, we have used a perturbative analysis based on DFT calculations of the internal energy of the disordered solid solution to examine atomic ordering tendencies in the $Ti_x$NbMoTaW and $Ti_x$VNbMoTaW refractory high-entropy alloys. From the perturbative analysis, we have fitted a pairwise, real-space interaction and explored the phase space further using Monte Carlo simulations. We have also discussed the origins of the dominant atom-atom correlations in terms of the materials' underlying electronic structure. In contrast to Chap. 4, where we used LDA-optimsed lattice parameters, in this chapter, we have used lattice parameters based on experimental measurements. For the NbMoTaW and VNbMoTaW systems, the results of this chapter are in good agreement with the previous chapter, confirming that our modelling approach is not highly sensitive to small changes in lattice parameter around the experimental value.

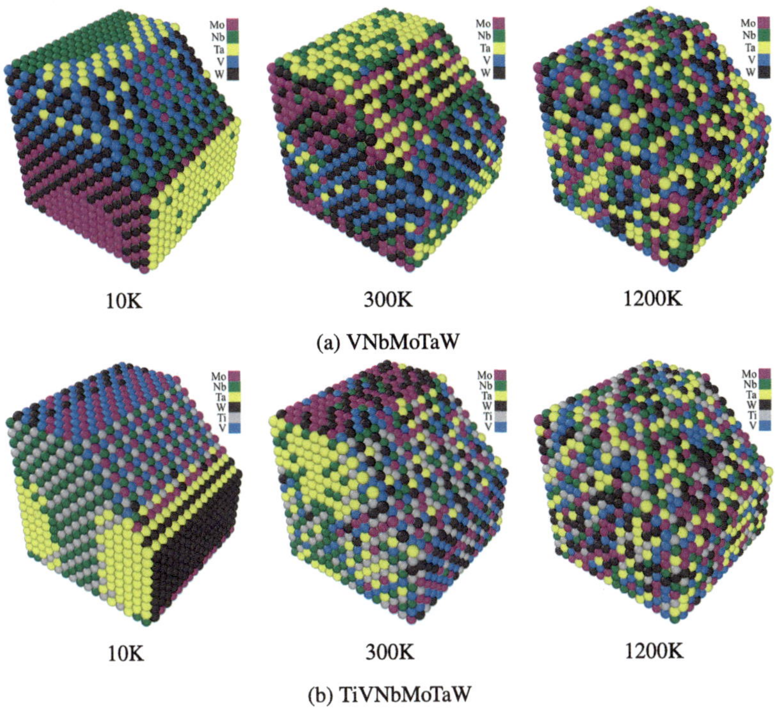


**Fig. 5.8** Visualisation of indicative equilibrium configurations for VNbMoTaW (*i.e.*, $Ti_x$VNbMoTaW, $x = 0$) and TiVNbMoTaW (*i.e.*, $Ti_x$VNbMoTaW, $x = 0$) at a selection of temperatures. Ti, V, Nb, Mo, Ta, and W are coloured grey, blue, green, pink, yellow, and black, respectively. A cut has been made through the simulation cell to make the configurations easier to interpret. Both the systems with and without Ti present end up becoming multiphase in our simulations as temperature decreases. Images generated using Ovito [16]

When Ti is not present (*i.e.*, the case $x = 0$) it is found that both NbMoTaW and VNbMoTaW form single-phase solid solutions down to relatively low temperatures, with predicted disorder-order transition temperatures of 557 K and 698 K respectively. The predicted transitions for these systems are chemical orderings, B2 and B32 respectively, and no significant phase segregation is expected. These results are consistent with both experimental and theoretical literature, as well with the results of the previous Chap. [10].

However, with increasing Ti concentration (*i.e.*, the case $x > 0$) it is found that strong atom-atom correlations emerge in the system. This leads to competition between phase segregation and phase ordering, with the perturbative analysis suggesting that Ti and Mo (and, to a lesser extent, V) tends to segregate from Nb, Ta, and W. This effect is amplified with increasing Ti concentration. This interpretation of the perturbative analysis is supported by analysis of our Monte Carlo simulations, where multiphase behaviour is clearly seen to emerge.

These results therefore shed light on the complex phase behaviour of these technologically relevant high-entropy alloys, as well as giving insight into the physical origins of the dominant atom-atom correlations in the solid solution. These results, and their agreement with experimental observations, also serve as further reassurance that this methodology is well-suited to studying the phase behaviour of multicomponent alloys.

## References

1. O. Senkov, G. Wilks, D. Miracle, C. Chuang, P. Liaw, Intermetallics **18**(9), 1758 (2010). https://doi.org/10.1016/j.intermet.2010.05.014
2. O. Senkov, G. Wilks, J. Scott, D. Miracle, Intermetallics **19**(5), 698 (2011). https://doi.org/10.1016/j.intermet.2011.01.004
3. Z. Han, N. Chen, S. Zhao, L. Fan, G. Yang, Y. Shao, K. Yao, Intermetallics **84**, 153 (2017) https://doi.org/10.1016/j.intermet.2017.01.007, https://linkinghub.elsevier.com/retrieve/pii/S0966979516305180
4. Z. Han, H. Luan, X. Liu, N. Chen, X. Li, Y. Shao, K. Yao, Mater. Sci. Eng. A **712**, 380 (2018). https://doi.org/10.1016/j.msea.2017.12.004
5. C.D. Woodgate, J.B. Staunton, J. Appl. Phys. **135**(13), 135106 (2024). https://doi.org/10.1063/5.0200862
6. M. Hoffmann, A. Ernst, W. Hergert, V.N. Antonov, W.A. Adeagbo, R.M. Geilhufe, H. Ben Hamed, Physica Status Solidi (b) **257**(7), 1900671 (2020). https://doi.org/10.1002/pssb.201900671
7. G.M. Stocks, W.M. Temmerman, B.L. Gyorffy, Phys. Rev. Lett. **41**(5), 339 (1978). https://doi.org/10.1103/PhysRevLett.41.339
8. J.P. Perdew, Y. Wang, Phys. Rev. B **45**(23), 13244 (1992). https://doi.org/10.1103/PhysRevB.45.13244
9. A.R. Denton, N.W. Ashcroft, Phys. Rev. A **43**(6), 3161 (1991). https://doi.org/10.1103/PhysRevA.43.3161
10. C.D. Woodgate, J.B. Staunton, Phys. Rev. Mater. **7**(1), 013801 (2023). https://doi.org/10.1103/PhysRevMaterials.7.013801
11. B. Zhang, M.C. Gao, Y. Zhang, S. Yang, S.M. Guo, Mater. Sci. Technol. **31**(10), 1207 (2015). https://doi.org/10.1179/1743284715Y.0000000031
12. W.P. Huhn, M. Widom, JOM **65**(12), 1772 (2013). https://doi.org/10.1007/s11837-013-0772-3
13. A. Fernández-Caballero, J.S. Wróbel, P.M. Mummery, D. Nguyen-Manh, J. Phase Equil. Diff. **38**(4), 391 (2017). https://doi.org/10.1007/s11669-017-0582-3
14. T. Kostiuchenko, F. Körmann, J. Neugebauer, A. Shapeev, Npj Comput. Mater. **5**(1), 55 (2019). https://doi.org/10.1038/s41524-019-0195-y
15. A.D. Kim, M. Widom, Phys. Rev. Mater. **7**(6), 063803 (2023). https://doi.org/10.1103/PhysRevMaterials.7.063803
16. A. Stukowski, Modell. Simul. Mater. Sci. Eng. **18**(1), 015012 (2010). https://doi.org/10.1088/0965-0393/18/1/015012
17. F. Körmann, A.V. Ruban, M.H. Sluiter, Mater. Res. Lett. **5**(1), 35 (2017). https://doi.org/10.1080/21663831.2016.1198837

# Chapter 6
# Atomic Short-Range Order and Phase Stability of the Cantor-Wu Medium- and High-Entropy Alloys

**Abstract** Following on from the earlier chapters on refractory high-entropy alloys, in which magnetism played no role, we now move on to studying alloys containing magnetic elements. Namely, we study the prototypical family of face centred cubic high-entropy alloys, the Cantor alloy (CrMnFeCoNi) and its derivatives, collectively referred to as the Cantor-Wu alloys. We study an indicative series of the Cantor-Wu alloys: NiCo, NiCoCr, NiCoFeCr, and NiCoFeMnCr. We find that atom-atom effective pair interactions are not approximated well either as pseudobinary or restricted to nearest-neighbour distance. Our computed order-disorder transition temperatures are low, consistent with experimental observations.We find that atomic ordering is dominated by correlations between Ni, Co, and Cr, while Fe and Mn interact weakly. Our linear-response analysis suggests $L1_2$ orderings in CrCoNi, CrFeCoNi, and CrMnFeCoNi which are all driven primarily by Cr atoms moving to one sublattice. Further atomistic modelling suggests that there is no true single-phase low-temperature ground state for these multicomponent systems. Instead, the single-phase solid solution is kept stable to low temperatures by the large configurational entropy and the Fe, Mn dilution effects.

## 6.1 Introduction

Following on from the results of the previous two chapters studying refractory HEAs, in which magnetism played no role, we now move on to consider what are arguably the prototypical family of HEAs, those forming on an fcc lattice, based on Ni, and containing other magnetic elements. This family consists of the Cantor alloy, CrMnFeCoNi [1], and its derivatives as classified by Wu *et al.* [2], collectively referred to as the Cantor-Wu alloys [3]. These are a family of two, three, four, and five component, equiatomic, fcc solid solutions, containing CoNi, FeNi, CrCoNi, MnCoNi, CrMnCoNi, and CrFeCoNi. Similarly to the refractory HEAs of Chaps. 4 and 5, they often have exceptional physical properties for applications, including remarkable hardness; excellent resistance to irradiation and fracture; and high tensile strength [4–7]. Although in principle one might expect any subset of the five elements of the Cantor alloy to form a solid solution, in practice this is not the case [2] and therefore



C. D. Woodgate, *Modelling Atomic Arrangements in Multicomponent Alloys*, Springer Series in Materials Science 346,
https://doi.org/10.1007/978-3-031-62021-8_6

a key challenge for theory is to understand the phase behaviour of these materials. In addition, there is extensive existing literature, both experimental [8–12] and computational [13–21], studying atomic arrangements in these systems with which to compare our results. We elect to study the three-component, medium-entropy CrCoNi, the four-component CrFeCoNi, and the five-component CrMnFeCoNi as a series of three multicomponent alloys with increasing configurational entropy.

As for the refractory HEAs of the previous chapters, here we once again apply the modelling approach developed in Chaps. 2 and 3 to these multicomponent systems, and make comparisons with experimental results and those obtained using alternative compuational techniques. Compared to the refractory systems, which are entirely non-magnetic, in these systems care must be taken to treat magnetism appropriately. The Curie temperatures of the majority of these systems are known to be well below room temperature [3]. It is therefore important to describe the paramagnetic state when modelling atomic arrangements, as this will be most reflective of the magnetic state of the material during processing.

This chapter is structured in a similar manner to Chap. 4. In Sect. 6.2 we will begin by examining the electronic structure of the atomically disordered solid solutions, to pull out features that will be relevant to atomic ordering in these materials. We will also examine selected binary subsystems for comparison, and to elucidate the electronic origins of ordering tendencies. Then, in Sect. 6.3, we will present results from the linear response theory and interpret our results in the context of the materials' electronic structure. As in Chap. 4, we will again illustrate the failure of the band-only approximation to successfully describe the phase behaviour of these systems. Proceeding, in Sects. 6.4 and 6.5 we fit to a pairwise Hamiltonian and perform atomistic modelling to further explore the phase space of these systems. We will also use the three component CrCoNi to illustrate why the pseudobinary approximation fails to successfully describe the phase behaviour of some multicomponent systems. Finally, in Sect. 6.6, the results are summarised and an outlook given.

## 6.2 Electronic Structure

As a starting point for linear response calculations, we again perform self-consistent DFT calculations to obtain the single-electron potentials and densities for the disordered alloys. These are computed assuming the materials are in a paramagnetic state. The treatment of the magnetic state as paramagnetic is well-justified; previous studies of magnetism in these materials have noted that all have low Curie temperatures, usually below room temperature and certainly below any temperature at which an alloy would be processed or annealed [3, 23–25]. Averaging over compositional disorder is achieved using the CPA [26], while the paramagnetic state is described within the disordered local moment (DLM) picture as outlined in Chap. 3 and [27]. (In this context, the DLM picture describes the average electronic structure of a system where the overall magnetisation is zero, *i.e.* above any magnetic ordering temperature.) In our calculations for these solid solutions on the fcc lattice, we find that only Fe and Mn support local moments, while Cr, Co, and Ni do not.

As for the refractory alloys of Chap. 4, our self-consistent calculations were performed using the *Hutsepot* code [28], with many of the same input and convergence parameters as for Chaps. 4 and 5. Spin-polarised, scalar-relativistic calculations were performed within the atomic sphere approximation (ASA) [29], using an angular momentum cutoff of $l_{\text{max}} = 3$ for basis set expansions, a $20 \times 20 \times 20$ **k**-point mesh for integrals over the Brillouin zone, and a 24 point semi-circular Gauss-Legendre grid in the complex plane to integrate over valence energies. We use the local density approximation and the exchange-correlation functional is that of Perdew-Wang [30]. Cubic fcc lattice parameters for the systems were set at the experimental values [1, 31, 32] of 3.53, 3.56, 3.57 and 3.59 Å for CoNi, CrCoNi, CrFeCoNi, and CrMnFeCoNi respectively. The additional CrNi and CrCo binary systems also had their lattice parameter set at 3.53 Å for consistency with the CoNi calculation.

Having performed a self-consistent calculation, we again plot the density of states (DoS) of the disordered alloys to gain an understanding of their electronic structure. Shown in Fig. 6.1 are plots of the total and species-resolved DoS for CrCoNi, CrFeCoNi, CrMnFeCoNi, as well as the three possible equiatomic binary subsystems of the ternary CrCoNi. The total DoS is given by the average of the species-resolved curves, weighted by their concentrations. It should be noted that, of the binaries, CoNi is expected to form a single-phase fcc solid solution, while CrNi and CrCo are not [33–35]. We force CrNi and CrCo onto the fcc lattice in our calculations to

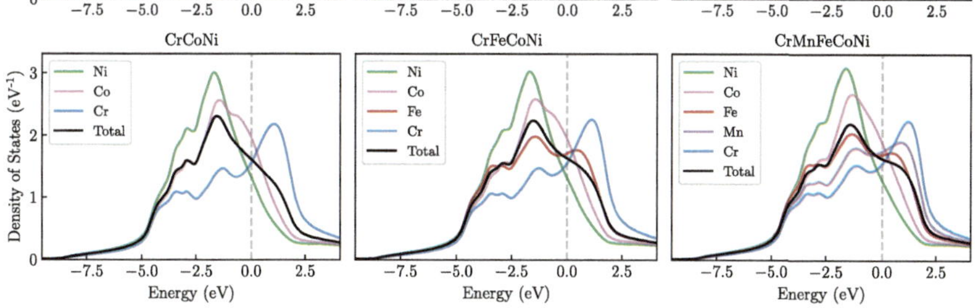


**Fig. 6.1** Comparison of the total and species-resolved DoS for the high temperature, disordered, paramagnetic states of the Cantor-Wu alloys considered in this work. The total DoS is given by the weighted average of the species-resolved curves. The Fermi energy for each system is marked as a grey dashed line. Compositional disorder is modelled via the CPA, while the paramagnetic state is described within the DLM picture. Reproduced from [22] with permission from the American Physical Society

better understand the differences between interactions between species when they are modelled in the binary setting as opposed to the multicomponent one.

We first consider the binary systems CrCo, CrNi, and CoNi. Despite the small valence difference between the two constituent elements, it is obvious from the DoS plots that CoNi will form a solid solution down to low temperatures, because of how similar the species resolved curves for Co and Ni are across the entire range of valence electron energies. In contrast, the two Cr-containing binaries have a Cr DoS curve with a peak lying above the Fermi energy, and far away from the peaks in either the Co or Ni curves, a consequence of the large valence differences between the species.

Perhaps the most notable feature of the figure is the difference between the CoNi binary and the ternary CrCoNi system; the addition of Cr completely alters the character of both the Ni and Co species-resolved curves, as well as that of the total DoS plotted in black. We suggest that, already, this provides evidence that the Ni-Co interaction in this system is likely to be poorly modelled as pseudobinary.

When the four component CrFeCoNi and the five component CrMnFeCoNi systems are considered, it can be seen that the species-resolved Fe and Mn curves lie close to the total DoS curves and do little to alter its shape. This suggests that these elements mix well with the other elements and dilute the solid solution. This is a consequence of Fe and Mn having a valence filling between that of the other three elements, and also of the magnetic disorder associated with these two elements in the DLM state.

## 6.3 Linear Response

Starting from the self-consistent potentials and charge densities, the linear response theory described in Chaps. 2 and 3 is used to calculate atomic short-range order from first principles and subsequently construct the chemical stability matrix. This calculation is performed in reciprocal space, and we sample a mesh of 56 points around the irreducible Brillioun zone, including high symmetry points. To visualise the results, we plot $S^{(2)}_{\alpha\alpha'}(\mathbf{k})$ along high-symmetry lines around the IBZ. In addition, we use the $S^{(2)}$ data to construct the chemical stability matrix around the IBZ at a temperature of $T = 1000$ K, and plot its eigenvalues, which will tell us about preferred atomic ordering. Plots of these data for all six considered systems are shown in Fig. 6.2. Eigenvalues, and their corresponding eigenvectors (or polarisations) in concentration space are tabulated for a few special points around the IBZ in Table 6.2.

Intriguingly, for Co-Ni pairs, the $S^{(2)}$ data (Fig. 6.2a) has a markedly different $\mathbf{k}$-dependence in the multicomponent systems as compared to the binary CoNi alloy. Again, we suggest that this indicates that the pseudobinary approximation[1] will fail to describe these systems successfully. In addition, we note that by far the most

[1] The pseudobinary approximation assumes that atom-atom interactions in a multicomponent system can be taken from the respective binary alloys.

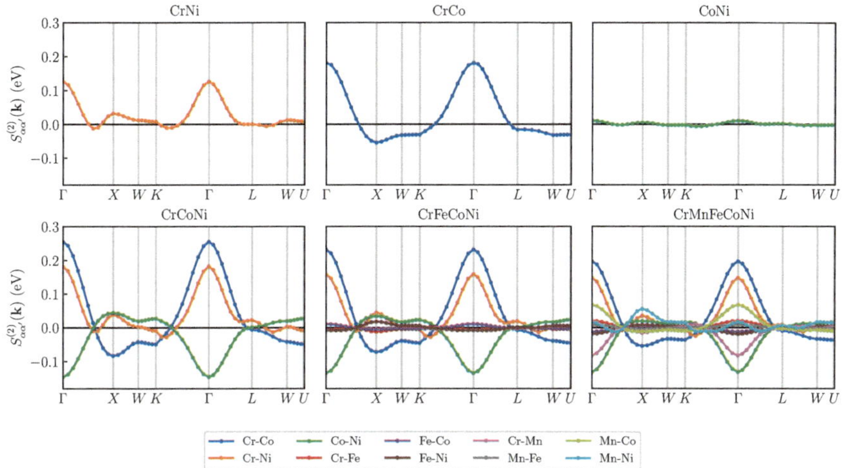


(a) $S^{(2)}_{\alpha\alpha'}(\mathbf{k})$

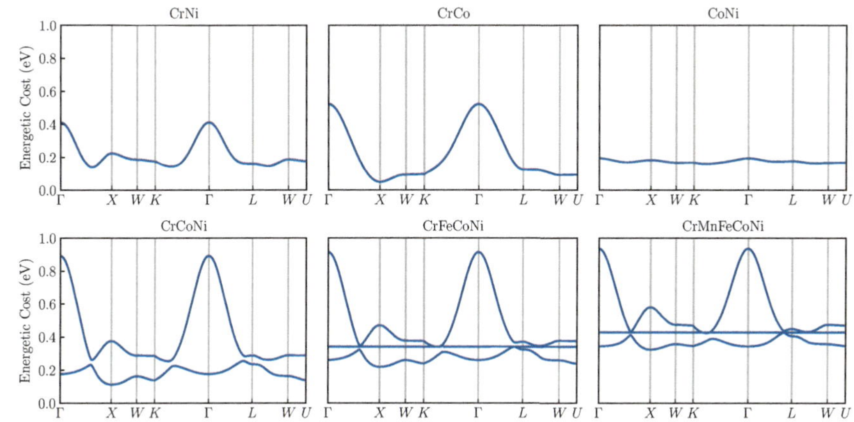


(b) Eigenvalues of $[\beta^{-1}\Psi^{-1}(\mathbf{k})]_{\alpha\alpha'} = \left[\beta^{-1}\frac{\delta_{\alpha\alpha'}}{c_\alpha} - S^{(2)}_{\alpha\alpha'}(\mathbf{k})\right]$ at $T$ = 1000 K

**Fig. 6.2** Plots of the raw $S^{(2)}_{\alpha\alpha'}(\mathbf{k})$ data (top) and eigenvalues of the chemical stability matrix (bottom), around the irreducible Brillouin Zone of the face centred cubic lattice for the for the considered Cantor-Wu systems. All eigenvalues are evaluated at a temperature of $T = 1000$ K. Of the binaries, CrCo demonstrates the strongest ordering tendencies, as its minimum eigenvalue is lowest-lying. It can be seen from the multicomponent plots that the addition of Fe and Mn does not significantly alter the shape of the dominant modes in CrCoNi. Instead, these two elements introduce two near-flat modes which, in reciprocal space, are polarised predominantly along the Fe and Mn directions, indicative of their diluting effect. The upward trend in the lowest lying eigenvalue from CrCoNi through to CrMnFeCoNi is due to the increasingly large contribution to the chemical stability matrix from the entropy of mixing. Reproduced from [22] with permission from the American Physical Society

**Table 6.1** Calculated transition temperatures for the considered Cantor-Wu systems, including the mode and eigenvector for which the instability occurs. We note that transition temperatures are computed using a mean field theory and so are expected to be overestimates of their true values. The three, four, and five component systems show a clear trend that increasing configurational entropy stabilises the solid solution and drives down the ordering temperature. Note that $\Delta c_{\alpha}$ for Fe and Mn is always small, indicative of weak correlations. Also note that the values 0.65 and 0.6 for CrNi and CoNi respectively are really expected to be 2/3, but this precise point was not sampled. The small numerical differences between these results and those of [22] are due to adjustments in the numerical solvers

| System | $T_{\text{us}}$ (K) | $\mathbf{k}_{\text{us}}$ | $\Delta c_{\text{Cr}}$ | $\Delta c_{\text{Mn}}$ | $\Delta c_{\text{Fe}}$ | $\Delta c_{\text{Co}}$ | $\Delta c_{\text{Ni}}$ |
|---|---|---|---|---|---|---|---|
| CrNi | 175 | (0, 0, 0.65) | 0.707 | | | | −0.707 |
| CrCo | 774 | (0, 0, 1) | 0.707 | | | −0.707 | |
| CoNi | 59 | (0, 0.6, 0.6) | | | | 0.707 | −0.707 |
| CrCoNi | 581 | (0, 0, 1) | 0.723 | | | −0.690 | −0.032 |
| CrFeCoNi | 380 | (0, 0, 1) | 0.722 | | −0.049 | −0.689 | 0.016 |
| CrMnFeCoNi | 258 | (0, 0, 1) | 0.722 | 0.010 | −0.082 | −0.686 | 0.035 |

important $S^{(2)}$ contributions come from the Cr-Co, Cr-Ni, and Co-Ni pairs, with contributions involving Mn and Fe being far weaker, suggesting these elements alloy well.

Proceeding, we consider the eigenvalues of the chemical stability matrix, Fig. 6.2b. Of the three binary systems, CrCo displays the strongest ordering tendencies, evidenced by the fact that its minimum eigenvalue is lowest-lying. This minimum occurs at the $X$ point, indicative of $\text{L}1_0$ or $\text{L}1_2$ ordering. By contrast, the exceptionally flat concentration wave mode associated with CoNi demonstrates that interactions between Ni and Co in that system are weak and the system is likely to remain a solid solution down to a much lower temperature than the other systems considered, in alignment with the $S^{(2)}$ data. The third binary system, CrNi, has competing minima along the lines $\Gamma$-$X$ and $K$-$\Gamma$. This is understood by noting that we are relatively close to the composition $Ni_2Cr$, which is known to form an intermetallic phase with an $MoPt_2$ crystal structure [36, 37], described by $\mathbf{k}_{\text{us}} = (0, 2/3, 2/3)$, a point along the line $K$-$\Gamma$.

When looking at the multicomponent systems, for CrCoNi the minimum eigenvalue can be seen to lie at the $X$ point, indicative of $\text{L}1_0$ or $\text{L}1_2$ ordering.[2] Perhaps the most striking feature of the three plots for the multicomponent systems is that the two dominant modes displayed in CrCoNi maintain their shape as additional elements Fe and Mn are added to form the four- and five-component systems. The addition of Fe and Mn has two effects. The first is to introduce near-flat concentration wave modes which, in concentration space, are polarised along the Fe and Mn 'directions', indicating that Fe and Mn interact weakly with the other species in the system. (This is evidenced by the chemical polarisations seen in Table 6.2.) The

[2] Recall that the $\text{L}1_0$ and $\text{L}1_2$ structures, visualised in Chap. 2, are described by a wavevector $\mathbf{k}_{\text{us}} = (0, 0, 1)$ and equivalent.

**Table 6.2** Eigenvalues of the chemical stability matrix for the multicomponent Cantor-Wu systems at selected special points of the IBZ evaluated at $T = 1000$ K, and their corresponding eigenvectors, or chemical polarisations, in concentration space

| Material | Point | Eigenvalue (eV) | $\Delta c_{Cr}$ | $\Delta c_{Mn}$ | $\Delta c_{Fe}$ | $\Delta c_{Co}$ | $\Delta c_{Ni}$ |
|---|---|---|---|---|---|---|---|
| CrCoNi | Γ | 0.181 | 0.083 | | | 0.662 | −0.745 |
| | | 0.932 | 0.812 | | | −0.478 | −0.335 |
| | $X$ | 0.108 | −0.723 | | | 0.690 | 0.033 |
| | | 0.374 | −0.380 | | | −0.436 | 0.816 |
| | $K$ | 0.134 | −0.790 | | | 0.574 | 0.216 |
| | | 0.292 | −0.207 | | | −0.581 | 0.787 |
| CrFeCoNi | Γ | 0.263 | 0.036 | | 0.167 | 0.588 | −0.791 |
| | | 0.344 | −0.313 | | 0.850 | −0.403 | −0.134 |
| | | 0.949 | 0.807 | | 0.013 | −0.493 | −0.327 |
| | $X$ | 0.213 | −0.722 | | 0.049 | 0.689 | −0.016 |
| | | 0.344 | −0.320 | | 0.849 | −0.399 | −0.130 |
| | | 0.471 | 0.354 | | 0.162 | 0.340 | −0.856 |
| | $K$ | 0.236 | −0.788 | | 0.017 | 0.588 | 0.183 |
| | | 0.345 | −0.316 | | 0.847 | −0.409 | −0.123 |
| | | 0.380 | −0.173 | | −0.178 | −0.487 | 0.838 |
| CrMnFeCoNi | Γ | 0.346 | −0.117 | 0.326 | 0.116 | 0.475 | −0.800 |
| | | 0.431 | −0.240 | −0.116 | 0.871 | −0.395 | −0.120 |
| | | 0.431 | 0.420 | −0.801 | 0.157 | 0.370 | −0.146 |
| | | 0.968 | 0.743 | 0.198 | −0.058 | −0.531 | −0.352 |
| | $X$ | 0.312 | −0.722 | −0.010 | 0.082 | 0.686 | −0.035 |
| | | 0.430 | 0.407 | −0.246 | −0.714 | 0.513 | 0.039 |
| | | 0.431 | 0.256 | −0.759 | 0.529 | 0.185 | −0.210 |
| | | 0.585 | 0.216 | 0.404 | 0.067 | 0.181 | −0.868 |
| | $K$ | 0.345 | −0.756 | −0.091 | 0.077 | 0.628 | 0.142 |
| | | 0.431 | 0.251 | −0.759 | 0.539 | 0.173 | −0.204 |
| | | 0.431 | −0.403 | 0.252 | 0.703 | −0.529 | −0.024 |
| | | 0.475 | −0.059 | −0.391 | −0.098 | −0.311 | 0.859 |

second effect is to shift the value of the lowest lying eigenvalue upwards, associated with the contribution to the chemical stability matrix from the entropy of mixing, which is larger for systems with a greater number of components. Taken together with the $S^{(2)}$ data, we interpret this as Fe and Mn diluting the Ni-Co-Cr interactions and weakening any ordering tendencies. This exemplifies the term 'high-entropy'; the introduction of Fe and Mn does not introduce strong atom-atom interactions but rather serves to increase the configurational entropy and drive down any atomic ordering temperature.

Table 6.1 then shows temperatures at which the lowest eigenvalue passes through zero and an absolute instability occurs. (It is at this temperature that we infer an order/disorder transition.) We tabulate the **k**-vector for which this occurs, and also its polarisation in concentration space. For convenience, we visualise the (partially) ordered structures predicted by the linear-response analysis in Fig. 6.3, specifying the (partial) occupancy of each non-equivalent lattice site. These ordered structures are obtained by allowing the predicted concentration wave to 'grow' until (at least) one sublattice has one chemical species whose occupancy reaches either 0 or 1. This is then the largest physically permitted chemical fluctuation consistent with the concentration wave's polarisation. For most of these systems, an experimental disorder-order transition temperature is not readily obtainable because kinetics at these low temperatures are sluggish. However, given that ordered phases are not observed experimentally, and that we predict low ordering temperatures, we conclude that our results are in agreement with experiment.

The three binary systems illustrate the contrasting relative strength of the Ni-Co, Ni-Cr, and Co-Cr interactions when constrained to an fcc lattice. The predicted modes for CrNi and CoNi are consistent with existing theoretical and experimental literature on ASRO in these systems [37–40]. For CoNi, both the associated energies and the transition temperature are very small, consistent with our interpretation of the DoS plots and also with the experimental literature [33]. The addition of Cr to form the CrCoNi ternary causes significant change in the eigenvalues and eigenvectors of the chemical stability matrix, again demonstrating the shortcomings of the pseudobinary approximation.

The modes for which the instability occurs for CrCo, CrCoNi, CrFeCoNi, and CrMnFeCoNi are all $\mathbf{k}_{\text{us}} = (0, 0, 1)$, with the dominant components of the eigenvector associated with Co and Cr. This is indicative of Co and Cr setting up long-range order in an $L1_0$ or $L1_2$ structure. $L1_0$ indicates alternating layers of Co-rich, Cr-depleted and Co-depleted, Cr-rich, while $L1_2$ denotes Cr sitting on the corners of the cubic fcc unit cell. As the concentrations of elements in these equiatomic systems

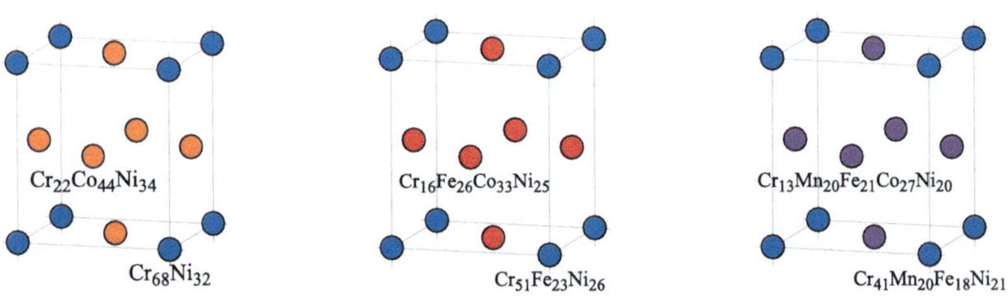


(a) CrCoNi (b) CrFeCoNi (c) CrMnFeCoNi

**Fig. 6.3** Visualisations of the partially ordered structures predicted by the perturbative stability analysis for CrCoNi (left) and CrFeCoNi (center), and CrMnFeCoNi (right) emerging at 581, 380 and 258 K respectively. All three orderings are L12 orderings driven primarily by Cr moving to the 'corners' of the fcc lattice. The partial occupancies of the two non-equivalent sites in each crystal structure are specified by the labels on the figures

are closer to 0.25 than 0.5, we infer $L1_2$ ordering. It can be seen from the table that the components of the eigenvectors associated with Ni, Fe, and Mn are all much smaller than the components associated with Co and Cr, meaning that the correlations involving Ni, Fe, and Mn are not the dominant ones. The earlier observation that the addition of Mn and Fe drives the lowest lying eigenvalue upwards is evidenced further by a reduction in the predicted transition temperature for the multicomponent systems with an increasing number of components, once again demonstrating how the entropy of mixing serves to stabilise the solid solution in these systems.

These results for the multicomponent systems compare favourably with existing literature. Experimentally, for the ternary CrCoNi, it is found that Co-Cr and Ni-Cr pairs are favoured, while Cr-Cr pairs are strongly disfavoured at nearest-neighbour distance [8, 9], which is consistent with our predicted $L1_2$ ordering. Similar results have been found by a number of previous computational studies, too [13, 16, 18, 19]. However, we note that some other computational studies have suggested a different, $MoPt_2$-like ordering occurring in this system [20, 21]. For the four component CrFeCoNi system, our results again broadly agree with existing computational studies, which find again that correlations between Co and Cr are dominant [13–15, 17]. One notable outlier is a study by Meshkov *et al.* [41] using a machine-learned interatomic potential to study the CrFeCoNi system. These authors predict, in agreement with our results, that Cr establishes preferentially on one sublattice, but suggest the temperature of this ordering is 1230°C (= 1503 K). We note that this high predicted ordering temperature is not in alignment with other theoretical studies, or with the experimentally observed solid solutions. Finally, for the five-component system, which is observed experimentally to form a single-phase solid solution [1], we note that our predicted $L1_2$ ordering is in alignment with a computational study by Mizuno *et al.* [42], which used Monte Carlo simulations with DFT energy evaluations to examine short-range order in this system.

### 6.3.1 The Failure of the Band-Only Approximation

As for the refractory HEAs of Chap. 4, it would not be unreasonable to ask whether the full linear-response theory outlined in Chap. 3 is necessary, given that the alloys in this chapter contain only $3d$ elements and there is significantly reduced charge transfer compared to the refractory systems. However, we find that failing to include the effects of charge-transfer and charge-response yields highly unphysical results. Visualised in Fig. 6.4 are the eigenvalues of the chemical stability matrix using the band-only approximation for each of the three multicomponent systems, with those computed using the full linear response theory plotted for comparison.

In all three cases, the minimum eigenvalue has shifted in location from $X$ to $\Gamma$, indicative of phase segregation. In addition, the band-only eigenvalues lie lower, indicative of higher ordering temperatures. For example, for CrCoNi, the predicted ordering temperature within the band-only approximation is 1225 K, with a wavevector $\mathbf{k}_{\text{us}} = (0, 0, 0)$ and a chemical polarisation indicating Cr phase segregating out of

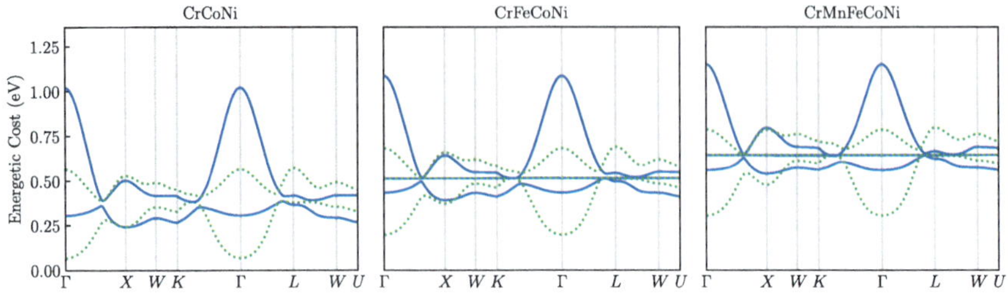


**Fig. 6.4** Comparison of the eigenvalues of the chemical stability matrix for the considered Cantor-Wu systems computed with the inclusion of charge transfer and charge response, versus those computed in the band-only approximation, both evaluated at $T = 1500$ K. Solid blue lines are results from a full $S^{(2)}$ calculation, while dashed green lines are band-only. The band-only approximation erroneously predicts phase segregation in all three systems, often at high temperatures

the solid solution. In CrFeCoNi, Cr is predicted to phase segregate out at a temperature of 923 K, while for the five-component CrMnFeCoNi, it is predicted that both Mn and Cr phase segregate out of the solid solution at a temperature of 790 K. All of these results are at odds with both the experimental and computational literatures for these systems, and we emphasise that it is therefore important to properly handle the effects of charge-transfer and charge-response when modelling these multicomponent alloys.

## 6.4 Pairwise Interactions

Having computed $S^{(2)}_{\alpha\alpha'}(\mathbf{k})$ in reciprocal space, we fit to a real-space effective pair interaction, writing $V^{(n)}_{\alpha\alpha'}$ for the interaction between species $\alpha$ and $\alpha'$ on shell $n$. This lattice-based model is well-justified for these systems, because it is known that lattice distortions due to varying atomic sizes are small [43]. It is necessary to choose a distance (and correspondingly a coordination shell, $n$) at which to truncate the fit.

Visualised in Fig. 6.5 are fitted $V^{(n)}_{\alpha\alpha'}$ to the first ten coordination shells of the fcc lattice. It can be seen that interactions are dominated by the first four coordination shells. Further analysis of the least-squares residuals of the fit as a function of number of shells included suggests that there is no significant improvement in fit given by including interactions beyond the fourth coordination shell for these systems. We therefore fit to the first four coordination shells to obtain our atom-atom interchange parameters (effective pair interactions) for use in atomistic simulations.

In Table 6.3 we tabulate these fitted effective pair interactions to fourth nearest neighbour distance for the three binaries and CrCoNi. For the four-component CrFeCoNi and the five-component CrMnFeCoNi these are provided in Appendix C, Table C.13. (The interactions for the binaries and the ternary are explicitly considered here to illustrate the failure of the pseudobinary approximation.) When the binary systems are considered, CoNi is by far the weakest interacting, consistent with our

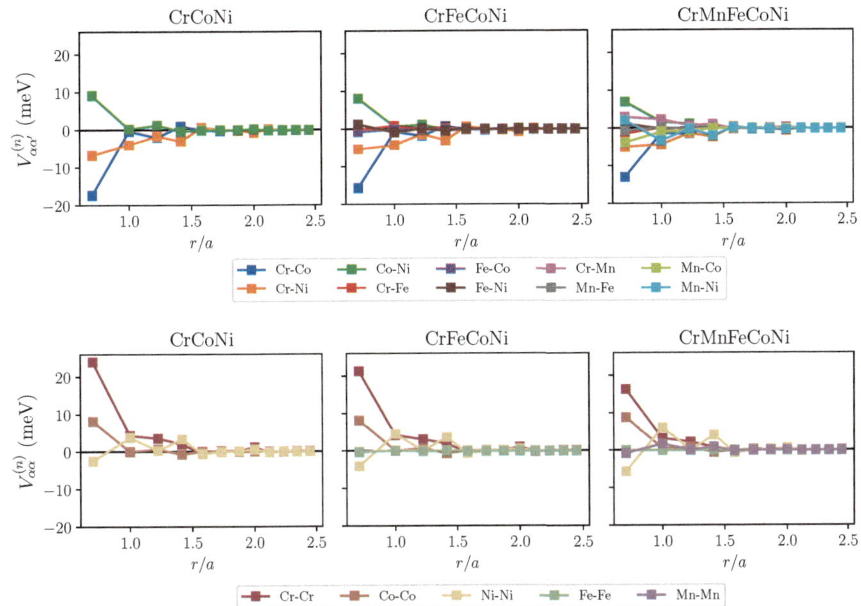


**Fig. 6.5** Visualisation of $V^{(n)}_{\alpha\alpha'}$ for the three multicomponent Cantor-Wu systems considered, fitted to the first ten coordination shells. It can be seen that interactions are dominated by the first four coordination shells

earlier observations. CrNi interacts more strongly, but by far the strongest interactions are found in CrCo, which we associate with constraining this system onto the fcc lattice. When the ternary CrCoNi system is considered, we see the clear failure of the pseudobinary approximation. Ni-Co interactions are an order of magnitude greater at nearest neighbour distance in the ternary CrCoNi as compared to the binary CoNi. Finally, when inspecting interactions for the four-component CrFeCoNi and five-component CrMnFeCoNi, we see that Fe and Mn interact weakly with all other elements in the system. This is further evidence that Fe and Mn have a diluting effect.

## 6.5 Atomistic Modelling

Using the obtained real-space interactions, we perform Monte Carlo simulations to better understand the nature of ordering, probe beyond any initial disorder-order transition, and visualise potential equilibrium configurations for the multicomponent systems CrCoNi, CrFeCoNi, and CrMnFeCoNi. All simulations were on-lattice and consisted of a system of $12 \times 12 \times 12$ cubic fcc unit cells, with 4 lattice sites per unit cell for a total of 6192 atoms in the simulation cell, with periodic boundary conditions applied. This size of supercell was selected as the largest cell reason-

**Table 6.3** $V_{\alpha\alpha'}^{(n)}$ for the considered binary Cantor-Wu systems, and CrCoNi, fitted to the first four coordination shells of the fcc lattice. All values in meV. A visual inspection shows that the Ni-Co interaction in the binary CoNi is completely different to the ternary CrCoNi, demonstrating the failure of the band-only approximation

| $V_{\alpha\alpha'}^{(1)}$ | Ni | Co | | $V_{\alpha\alpha'}^{(2)}$ | Ni | Co | |
|---|---|---|---|---|---|---|---|
| Ni | 0.3 | −0.3 | | Ni | 0.4 | −0.4 | |
| Co | −0.3 | 0.3 | | Co | −0.4 | 0.4 | |
| $V_{\alpha\alpha'}^{(3)}$ | Ni | Co | | $V_{\alpha\alpha'}^{(4)}$ | Ni | Co | |
| Ni | 0.0 | 0.0 | | Ni | 0.5 | −0.5 | |
| Co | 0.0 | 0.0 | | Co | −0.5 | 0.5 | |
| $V_{\alpha\alpha'}^{(1)}$ | Ni | Cr | | $V_{\alpha\alpha'}^{(2)}$ | Ni | Cr | |
| Ni | 2.7 | −2.7 | | Ni | 4.3 | −4.3 | |
| Cr | −2.7 | 2.7 | | Cr | −4.3 | 4.3 | |
| $V_{\alpha\alpha'}^{(3)}$ | Ni | Cr | | $V_{\alpha\alpha'}^{(4)}$ | Ni | Cr | |
| Ni | 1.7 | −1.7 | | Ni | 1.7 | −1.7 | |
| Cr | −1.7 | 1.7 | | Cr | −1.7 | 1.7 | |
| $V_{\alpha\alpha'}^{(1)}$ | Co | Cr | | $V_{\alpha\alpha'}^{(2)}$ | Co | Cr | |
| Co | 14.0 | −14.0 | | Co | 1.5 | −1.5 | |
| Cr | −14.0 | 14.0 | | Cr | −1.5 | 1.5 | |
| $V_{\alpha\alpha'}^{(3)}$ | Co | Cr | $V_{\alpha\alpha'}^{(4)}$ | Co | Cr | | |
| Co | 1.4 | −1.4 | Co | −0.9 | 0.9 | | |
| Cr | −1.4 | 1.4 | Cr | 0.9 | −0.9 | | |
| $V_{\alpha\alpha'}^{(1)}$ | Ni | Co | Cr | $V_{\alpha\alpha'}^{(2)}$ | Ni | Co | Cr |
| Ni | −3.0 | 9.3 | −6.3 | Cr | 3.6 | 0.4 | −4.0 |
| Co | 9.3 | 9.2 | −18.4 | Co | 0.4 | 0.0 | −0.5 |
| Cr | −6.3 | −18.4 | 24.7 | Ni | −4.0 | −0.5 | 4.5 |
| $V_{\alpha\alpha'}^{(3)}$ | Ni | Co | Cr | $V_{\alpha\alpha'}^{(4)}$ | Ni | Co | Cr |
| Ni | 0.3 | 1.3 | −1.6 | Cr | 3.3 | −0.1 | −3.2 |
| Co | 1.3 | 0.7 | −2.0 | Co | −0.1 | −0.7 | 0.8 |
| Cr | −1.6 | −2.0 | 3.7 | Ni | −3.2 | 0.8 | 2.4 |

ably computationally affordable. The initial conditions for each simulation were a random arrangement of atoms, and then the Metropolis algorithm with the so-called Kawasaki dynamics to conserve overall concentrations of each species was used. The systems were annealed from 1200 K in steps of 1 K, with $10^4$ steps per atom at each temperature to ensure equilibrium was achieved. At each temperature, a measure of the SHC was extracted, and the Warren-Cowley ASRO parameters were calculated based on equilibrated configurations at that temperature. An ensemble of sixteen simulations was used for each system and the results averaged, enabling extraction of an uncertainty estimate on the simulation data.

Figure 6.6 shows SHC curves and Warren-Cowley ASRO parameters on first- and second-nearest neighbour shells as a function of temperature for each system. For all three systems, the dominant correlations are similar and involve Ni, Co, and Cr. At nearest neighbour distance, Co-Cr pairs are strongly favoured, while Ni-Co and Cr-Cr pairs are disfavoured. At second nearest neighbour distance, Co-Cr pairs are disfavoured. This is indicative of an $L1_2$-like ordering, with Cr sitting on the corners of the fcc unit cell where possible. These results are in good agreement with the earlier linear response analysis.

As discussed in Sect. 6.3, existing experimental literature on the ternary CrCoNi system [8, 9] has found that Co-Cr and Ni-Cr nearest-neighbour pairs are favoured,

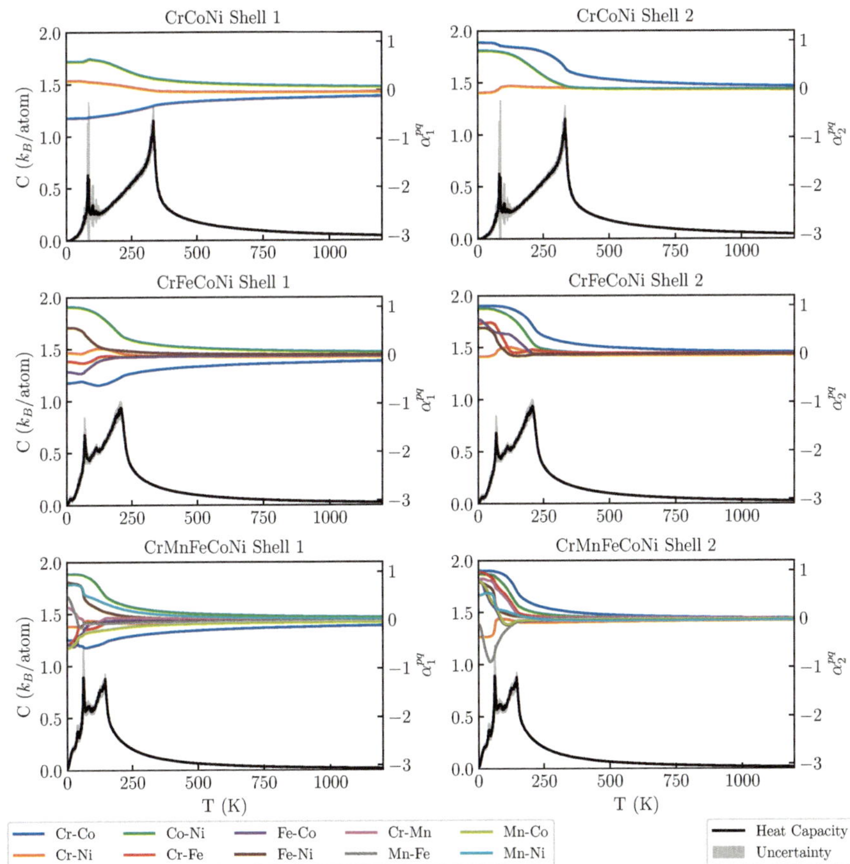


**Fig. 6.6** Plots of the specific heat capacity (SHC) and Warren-Cowley ASRO parameters as a function of temperature from Monte Carlo simulations of CrCoNi, CrFeCoNi, and CrMnFeCoNi, with energies evaluated using our computed effective pair interactions for each composition. Quantities are averaged across an ensemble of 16 simulations

while Cr-Cr neighbours are strongly disfavoured, and this is also found in computational studies employing DFT-based Monte Carlo simulations [13, 16].

Literature on the four component CrFeCoNi is limited to computational studies, owing to the lower ordering temperature making ASRO more difficult to observe experimentally. Tamm *et al.* [13] found that Cr-Cr pairs were disfavoured, while Ni-Cr and Co-Cr pairings were favoured, in agreement with our work. The same work suggests that Cr atoms in this material begin to form a simple cubic lattice, *i.e.* the $L1_2$ structure, which is consistent with our linear response analysis and existing literature [14, 15, 17, 44]. However, Tamm *et al.* also suggested that the number of Ni-Ni pairs decreases (we find an increase) and that there is significant ASRO between Fe and the other three elements, while all our results show Fe interacting weakly and demonstrating little to no tendency to order. We believe that this disagreement arises because the work models the system in a ferromagnetic (FM) state, while we model the paramagnetic (PM) state. This disagreement will be addressed explicitly in Chap. 7, where we will model these systems in a magnetically ordered state. Fe in particular has a large local moment associated with it, and we expect the nature of predicted order in this material to be different between FM and PM states. We model the PM state because the Curie temperature for these alloys is typically of the order of 100 K [3, 17, 45], well below the temperatures at which these materials are processed and annealed experimentally [32]. Another potential source of the discrepancy is that the Tamm study employed lattice MC simulations with energies evaluated *ab initio*, so was limited to less than 20 MC steps per atom. By contrast, our extracted atomistic modelling parameters enable us to comfortably achieve $10^4$ MC steps per atom, which is the expected number needed to achieve equilibrium and enables us to interpret our results with confidence. We also note that another first principles study by Schönfeld *et al.* found similar results to ours; Co-Cr and Ni-Cr correlations were dominant, with Fe exhibiting little to no correlation with other elements [17].

Finally, for the quinary CrMnFeCoNi, there is little literature on ASRO with which to compare, both experimental and computational. The lack of experimental data is associated with the low atomic ordering temperature, while the lack of computational results is largely associated with the fact that most supercell-based techniques are challenged by the huge space of possible configurations for a five-component alloy. Our results verify that any order in this material is likely to develop at such low temperatures as to be experimentally inaccessible; we find the pair distributions to be very close to uniform down to almost room temperature. However, we note that our predicted ASRO is in agreement with the computational study of [42], which used Monte Carlo simulations with energies evaluated directly via DFT calculations. In addition, our eventual predicted phase decomposition is broadly in alignment with [47], which found clustering of elements following simulated annealing using a modified embedded atom potential.

Studying the low-temperature behaviour predicted within our modelling, for all three multicomponent systems at lower temperatures, we observe that the system forms a large number of Ni-Ni nearest neighbour pairs which, in combination with visual analysis of the generated configurations shown in Fig. 6.7, we associate with

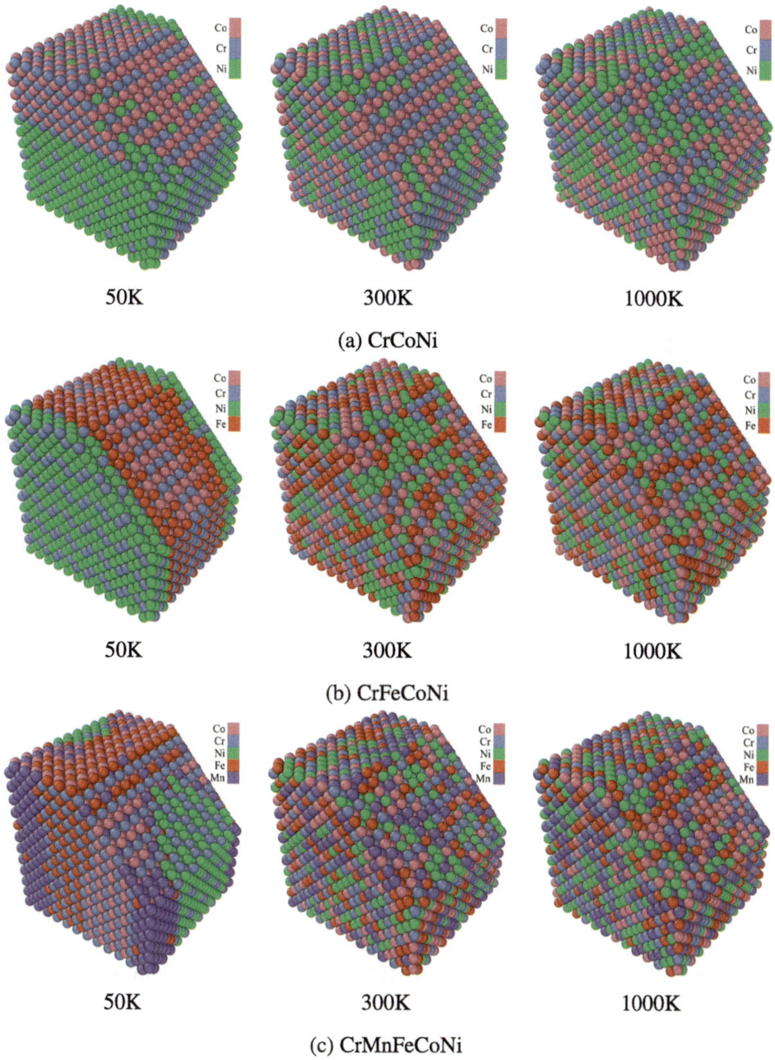


**Fig. 6.7** Visualised equilibrium configurations from Monte Carlo simulations using the effective pair interaction for the three considered multicomponent Cantor-Wu systems at temperatures of 1000, 300 and 50 K. Cr, Mn, Fe, Co, and Ni are coloured blue, purple, red, pink, and green respectively. A cut has been made through the simulation cell to make atomic ordering more clearly visible. Images generated using Ovito [46]

the system phase segregating into an Ni-rich region and another region which is Ni-deficient. The temperature at which this occurs in our simulations is typically below room temperature so unlikely to be experimentally accessible, but nonetheless this weakens the case for any study searching for a single-phase, 0 K ground state.

## 6.6 Summary

In summary, we have predicted the dominant ASRO and associated disorder-order transition temperatures for a subset of the Cantor-Wu high entropy alloys, and elucidated qualitative aspects of its origins in terms of the electronic structure of the disordered solid solution. Further, we took our reciprocal space results and mapped back to pairwise interchange parameters suitable for atomistic modelling. Lattice-based Monte Carlo simulations using these parameters validated the linear response analysis and gave a clear, visual description of the nature of order in these materials.

Our results are in excellent agreement with existing literature, both theoretical and experimental. Further, we are able to highlight problems with some existing studies and explain why Fe and Mn display extremely weak ordering tendencies in these materials, going beyond the state of the art. We are also able to achieve this using a fraction of the computational resources of previous studies, owing to the efficiency of our effective medium theory and the simplicity of our Hamiltonian used for atomistic modelling, as highlighted in Chap. 4.

We take this as evidence that our approach reliably describes the nature of order in multicomponent alloys containing magnetic elements, and that it can therefore be used to study other multicomponent systems to elucidate the nature of atomic ordering.

## References


1. B. Cantor, I.T.H. Chang, P. Knight, A.J.B. Vincent, Mater. Sci. Eng.: A **375–377**, 213 (2004). https://doi.org/10.1016/j.msea.2003.10.257
2. Z. Wu, H. Bei, F. Otto, G.M. Pharr, E.P. George, Intermetallics **46**, 131 (2014). https://doi.org/10.1016/j.intermet.2013.10.024
3. D. Billington, A.D.N. James, E.I. Harris-Lee, D.A. Lagos, D. O'Neill, N. Tsuda, K. Toyoki, Y. Kotani, T. Nakamura, H. Bei, S. Mu, G.D. Samolyuk, G.M. Stocks, J.A. Duffy, J.W. Taylor, S.R. Giblin, S.B. Dugdale, Phys. Rev. B **102**(17), 174405 (2020). https://doi.org/10.1103/PhysRevB.102.174405
4. M.H. Tsai, Entropy **15**(12), 5338 (2013). https://doi.org/10.3390/e15125338
5. B. Gludovatz, A. Hohenwarter, D. Catoor, E.H. Chang, E.P. George, R.O. Ritchie, Science **345**(6201), 1153 (2014). https://doi.org/10.1126/science.1254581
6. Y. Zhang, T.T. Zuo, Z. Tang, M.C. Gao, K.A. Dahmen, P.K. Liaw, Z.P. Lu, Prog. Mater. Sci. **61**, 1 (2014). https://doi.org/10.1016/j.pmatsci.2013.10.001
7. E.J. Pickering, N.G. Jones, Int. Mater. Rev. **61**(3), 183 (2016). https://doi.org/10.1080/09506608.2016.1180020
8. F. Zhang, S. Zhao, K. Jin, H. Xue, G. Velisa, H. Bei, R. Huang, J. Ko, D. Pagan, J. Neuefeind, W. Weber, Y. Zhang, Phys. Rev. Lett. **118**(20), 205501 (2017). https://doi.org/10.1103/PhysRevLett.118.205501
9. R. Zhang, S. Zhao, J. Ding, Y. Chong, T. Jia, C. Ophus, M. Asta, R.O. Ritchie, A.M. Minor, Nature **581**(7808), 283 (2020). https://doi.org/10.1038/s41586-020-2275-z
10. H.C. Robarts, T.E. Millichamp, D.A. Lagos, J. Laverock, D. Billington, J.A. Duffy, D. O'Neill, S.R. Giblin, J.W. Taylor, G. Kontrym-Sznajd, M. Samsel-Czeka la, H. Bei, S. Mu, G.D. Samolyuk, G.M. Stocks, S.B. Dugdale, Phys. Rev. Lett. **124**(4), 046402 (2020). https://doi.org/10.1103/PhysRevLett.124.046402

11. L. Zhou, Q. Wang, J. Wang, X. Chen, P. Jiang, H. Zhou, F. Yuan, X. Wu, Z. Cheng, E. Ma, Acta Materialia **224**, 117490 (2022). https://doi.org/10.1016/j.actamat.2021.117490
12. L. Li, Z. Chen, S. Kuroiwa, M. Ito, K. Yuge, K. Kishida, H. Tanimoto, Y. Yu, H. Inui, E.P. George, Acta Materialia **243**, 118537 (2023). https://doi.org/10.1016/j.actamat.2022.118537
13. A. Tamm, A. Aabloo, M. Klintenberg, M. Stocks, A. Caro, Acta Materialia **99**, 307 (2015). https://doi.org/10.1016/j.actamat.2015.08.015
14. C. Niu, A.J. Zaddach, A.A. Oni, X. Sang, J.W. Hurt, J.M. LeBeau, C.C. Koch, D.L. Irving, Appl. Phys. Lett. **106**(16), 161906 (2015). https://doi.org/10.1063/1.4918996
15. T. Fukushima, H. Katayama-Yoshida, K. Sato, M. Ogura, R. Zeller, P.H. Dederichs, J. Phys. Soc. Jpn. **86**(11), 114704 (2017). https://doi.org/10.7566/JPSJ.86.114704
16. J. Ding, Q. Yu, M. Asta, R.O. Ritchie, Proc. Natl. Acad. Sci. **115**(36), 8919 (2018). https://doi.org/10.1073/pnas.1808660115
17. B. Schönfeld, C.R. Sax, J. Zemp, M. Engelke, P. Boesecke, T. Kresse, T. Boll, T. Al-Kassab, O.E. Peil, A.V. Ruban, Phys. Rev. B **99**(1), 014206 (2019). https://doi.org/10.1103/PhysRevB.99.014206
18. Z. Pei, R. Li, M.C. Gao, G.M. Stocks, npj Comput. Mater. **6**(1), 122 (2020). https://doi.org/10.1038/s41524-020-00389-1
19. F. Walsh, M. Asta, R.O. Ritchie, Proc. Natl. Acad. Sci. **118**(13), e2020540118 (2021). https://doi.org/10.1073/pnas.2020540118
20. J.P. Du, P. Yu, S. Shinzato, F.S. Meng, Y. Sato, Y. Li, Y. Fan, S. Ogata, Acta Materialia **240**, 118314 (2022). https://doi.org/10.1016/j.actamat.2022.118314
21. S. Ghosh, V. Sotskov, A.V. Shapeev, J. Neugebauer, F. Körmann, Phys. Rev. Mater. **6**(11), 113804 (2022). https://doi.org/10.1103/PhysRevMaterials.6.113804
22. C.D. Woodgate, J.B. Staunton, Phys. Rev. B **105**(11), 115124 (2022). https://doi.org/10.1103/PhysRevB.105.115124
23. B.C. Sales, K. Jin, H. Bei, G.M. Stocks, G.D. Samolyuk, A.F. May, M.A. McGuire, Sci. Rep. **6**(1), 26179 (2016). https://doi.org/10.1038/srep26179
24. Y.F. Kao, S.K. Chen, T.J. Chen, P.C. Chu, J.W. Yeh, S.J. Lin, J. Alloys Compd. **509**(5), 1607 (2011). https://doi.org/10.1016/j.jallcom.2010.10.210
25. O. Schneeweiss, M. Friák, M. Dudová, D. Holec, M. Šob, D. Kriegner, V. Holý, P. Beran, E.P. George, J. Neugebauer, A. Dlouhý, Phys. Rev. B **96**(1), 014437 (2017). https://doi.org/10.1103/PhysRevB.96.014437
26. J.S. Faulkner, G.M. Stocks, Y. Wang, *Multiple Scattering Theory: Electronic Structure of Solids*, 1st edn. (IOP Publishing, Bristol, UK, 2018)
27. B.L. Gyorffy, A.J. Pindor, J. Staunton, G.M. Stocks, H. Winter, J. Phys. F: Metal Phys. **15**(6), 1337 (1985). https://doi.org/10.1088/0305-4608/15/6/018
28. M. Hoffmann, A. Ernst, W. Hergert, V.N. Antonov, W.A. Adeagbo, R.M. Geilhufe, H. Ben Hamed, Phys. Status Solidi (b) **257**(7), 1900671 (2020). https://doi.org/10.1002/pssb.201900671
29. G.M. Stocks, W.M. Temmerman, B.L. Gyorffy, Phys. Rev. Lett. **41**(5), 339 (1978). https://doi.org/10.1103/PhysRevLett.41.339
30. J.P. Perdew, Y. Wang, Phys. Rev. B **45**(23), 13244 (1992). https://doi.org/10.1103/PhysRevB.45.13244
31. Y. Zhang, G.M. Stocks, K. Jin, C. Lu, H. Bei, B.C. Sales, L. Wang, L.K. Béland, R.E. Stoller, G.D. Samolyuk, M. Caro, A. Caro, W.J. Weber, Nat. Commun. **6**(1), 8736 (2015). https://doi.org/10.1038/ncomms9736
32. B. Yin, S. Yoshida, N. Tsuji, W.A. Curtin, Nat. Commun. **11**(1), 2507 (2020). https://doi.org/10.1038/s41467-020-16083-1
33. T. Nishizawa, K. Ishida, Bull. Alloy Phase Diagr. **4**(4), 390 (1983). https://doi.org/10.1007/BF02868090
34. P. Nash, Bull. Alloy Phase Diagr. **7**(5), 466 (1986). https://doi.org/10.1007/BF02867812
35. K. Ishida, T. Nishizawa, Bull. Alloy Phase Diagr. **11**(4), 357 (1990). https://doi.org/10.1007/BF02843315

36. J.B. Staunton, D.D. Johnson, F.J. Pinski, Phys. Rev. B **50**(3), 1450 (1994). https://doi.org/10.1103/PhysRevB.50.1450
37. M. Rahaman, B. Johansson, A.V. Ruban, Phys. Rev. B **89**(6), 064103 (2014). https://doi.org/10.1103/PhysRevB.89.064103
38. M. Hirabayashi, M. Koiwa, K. Tanaka, T. Tadaki, T. Saburi, S. Nenno, H. Nishiyama, Trans. Jpn. Inst. Metals **10**(5), 365 (1969). https://doi.org/10.2320/matertrans1960.10.365
39. B. Schönfeld, L. Reinhard, G. Kostorz, W. Bührer, Phys. Status Solidi (b) **148**(2), 457 (1988). https://doi.org/10.1002/pssb.2221480203
40. W. Schweika, H.G. Haubold, Phys. Rev. B **37**(16), 9240 (1988). https://doi.org/10.1103/PhysRevB.37.9240
41. E. Meshkov, I. Novoselov, A. Shapeev, A. Yanilkin, Intermetallics **112**, 106542 (2019). https://doi.org/10.1016/j.intermet.2019.106542
42. M. Mizuno, K. Sugita, H. Araki, Results Phys. **34**, 105285 (2022). https://doi.org/10.1016/j.rinp.2022.105285
43. H.S. Oh, D. Ma, G.P. Leyson, B. Grabowski, E.S. Park, F. Körmann, D. Raabe, Entropy **18**(9), 321 (2016). https://doi.org/10.3390/e18090321
44. Q. He, P. Tang, H. Chen, S. Lan, J. Wang, J. Luan, M. Du, Y. Liu, C. Liu, C. Pao, Y. Yang, Acta Materialia **216**, 117140 (2021). https://doi.org/10.1016/j.actamat.2021.117140
45. M.S. Lucas, D. Belyea, C. Bauer, N. Bryant, E. Michel, Z. Turgut, S.O. Leontsev, J. Horwath, S.L. Semiatin, M.E. McHenry, C.W. Miller, J. Appl. Phys. **113**(17), 17A923 (2013). https://doi.org/10.1063/1.4798340
46. A. Stukowski, Model. Simul. Mater. Sci. Eng. **18**(1), 015012 (2010). https://doi.org/10.1088/0965-0393/18/1/015012
47. I.N. Kar'kin, L.E. Kar'kina, Y.N. Gornostyrev, Phys. Metals Metall. **124**(10), 1026 (2023). https://doi.org/10.1134/S0031918X23601671

# Chapter 7
# A Cautionary Tale: Treatment of the Magnetic State in the Cantor-Wu Alloys

**Abstract** Previously, we examined the nature of atomic short-range order and the overall phase stability of the high-entropy Cantor alloy, CrMnFeCoNi, and its derivatives when modelled in their paramagnetic states within the disordered local moment (DLM) picture, which we believe to be reflective of the electronic structure of these materials at typical processing temperatures. However, it is also possible to model these systems in magnetically ordered (ferrimagnetic) and non-magnetic states. In this chapter, we examine the consequences of differing treatements of magnetism on predicted atomic ordering tendencies. Our results shed light on competition between $MoPt_2$ and $L1_2$ ordering in CrCoNi, as well as on how the strength of interactions between Fe, Mn, and other elements depends on how magnetism is treated. These results also have experimental implications, as they suggest it may be possible to tune the nature of atomic arrangements in some multicomponent alloys by applying a magnetic field during the annealing process.

## 7.1 Introduction

In the previous chapter, we modelled the Cantor alloy and its derivatives in their paramagnetic state using the DLM picture. This is because we believe that this is reflective of the magnetic state of these systems at typical materials processing temperatures, which are well above room temperature. However, a paramagnetic state, one in which atoms have magnetic moments associated with them but the system still has overall zero magnetisation, is computationally challenging to model within conventional density functional theory calculations using supercells. It is possible to perform calculations with non-collinear spins [1], but these are computationally expensive, so many studies return to a conventional collinear spin calculation. Alternatively, it is possible to perform calculations in which there is no spin polarisation.

In this chapter, within the developed computational framework, we will compare the effect of different treatments of alloys' magnetic states on predicted atomic arrangements. We choose to study the systems CrCoNi, CrFeCoNi, and CrMnFeCoNi, *i.e.*, the multicomponent systems of Chap. 6. We will see that there are profound differences in predicted atomic arrangements based on whether these materials are



C. D. Woodgate, *Modelling Atomic Arrangements in Multicomponent Alloys*, Springer Series in Materials Science 346,
https://doi.org/10.1007/978-3-031-62021-8_7

modelled as non-magnetic, paramagnetic, or magnetically ordered. These results have implications for materials modelling, but also for materials processing, as they suggest that spin-polarising the electronic structure of a material, *e.g.*, by applying a magnetic field during the annealing process, could alter how atoms arrange themselves in a material. This could, in turn, alter macroscopic materials properties.

In the context of high-entropy alloys, treatment of magnetism is understood to impact the results of materials modelling. Studies by Tamm *et al.* [2] and Ding *et al.* [3] used lattice-based Monte Carlo simulations with energies evaluated via DFT to study atomic arrangements in CrCoNi and CrFeCoNi. These calculations employed a spin-polarised scheme in which atoms are permitted to acquire magnetic moments and magnetically order, reflective of material behaviour at low temperatures. Indeed, in the previous chapter, we suggested that this factor could account for the discrepancy between our results (modelling the paramagnetic state) and theirs. This is justified by Ding *et al.*, who noted that the specific ASRO predicted differed depending on whether or not spin-polarisation was invoked in their calculations, although no results for the unpolarised calculations are provided. A later study by Walsh *et al.* [4] suggested that ASRO in CrCoNi is magnetically driven, and other studies by Niu *et al.* [5] and Fukushima *et al.* [6] on CrFeCoNi suggested that atomic ordering in this system is also driven by magnetism. By contrast, Ghosh *et al.* [7] found that their results modelling atomic short-range order in the CrCoNi system were unaffected by the treatment of magnetism in the system.

On a more practical, experimental level, it is known that annealing some alloys in an applied magnetic field can influence atomic arrangements and, therefore, materials properties. For example, in the late 1800s and early 1900s it was shown that annealing steel under applied static or alternating magnetic field conditions can influence its magnetic properties [8, 9]. Another example is given by the technologically significant Alnico magnets, where the application of a magnetic field during heat treatment can yield improved magnetic properties [10]. Further examples where magnetic field annealing can affect materials properties include soft and ultrasoft nanocrystalline materials such as Fe-Zr-B [11] and Fe-Si-B-M [12]. We note however that, at present, magnetic field annealing of medium- and high-entropy alloys is an unexplored area of research.

This chapter is structured as follows. In Sect. 7.2, we outline a key example of why treatment of the magnetic state matters by comparing the energy scales associated with chemical and magnetic ordering for the binary FeNi alloy. Then, in Sect. 7.3, we compare the results of self-consistent DFT calculations for the considered multicomponent systems in the three different magnetic states outlined above, and highlight how treatment of magnetism impacts the electronic structure of these materials. We then present results from the linear response theory applied to the self-consistent potentials describing these three magnetic states for each system. In Sects. 7.5 and 7.6, we fit the data from each set of linear response calculations to the pairwise model of earlier chapters and perform Monte Carlo simulations to give further insight into the impact of treatment of magnetism on the predicted atomic short-range order and phase stability of these systems. Finally, in Sect. 7.7, we summarise the results and give an outlook on their implications.

## 7.2 Energy Scales: Atomic Ordering Versus Magnetic Ordering

As an illustrative example of why care should be taken over treatment of a material's magnetic state, consider first the binary alloy FeNi, which is known to order in the $L1_0$ structure at low temperatures [13], but forms an fcc solid solution (A1 structure) at higher temperatures.

Using *Hutsepot* and the ASA, we perform four self-consistent calculations to obtain the DFT energy (per atom) of the system in six different states:

1. The A1 structure in a non-magnetic state.
2. The A1 structure in the paramagnetic (DLM) state.
3. The A1 structure in the ferromagnetic state.
4. The $L1_0$ structure in a non-magnetic state.
5. The $L1_0$ structure in the paramagnetic (DLM) state.
6. The $L1_0$ structure in the ferromagnetic state.

All calculations are performed using the body-centred tetragonal representation of the fcc lattice, with a two-atom basis, which enables the $L1_0$ and A1 structures to be described in unit cells with identical geometry, meaning energy differences come purely from magnetic and compositional order/disorder. Using the energy-per atom of the paramagnetic, A1 calculation as the zero of the energy scale, we compute energy differences between the described states and tabulate them in Table 7.1.

It can be seen that the energy scale associated with magnetic ordering is of the same order of magnitude as that associated with chemical ordering, and care should therefore be taken to ensure the magnetic state is treated appropriately. Magnetic interactions in binary Fe-Ni alloys are likely to be more important than in the Cantor-Wu systems, but the principle remains the same: care should be taken to treat magnetism correctly.

**Table 7.1** Comparison of the relative energy scales of chemical and magnetic ordering for $L1_0$ FeNi. Note that the total DFT energy of the paramagnetic A1 calculation is set as the zero of the energy scale. Note also that the DFT energy of the paramagnetic state is an *ensemble average*

| Atomic arrangement | Magnetic state | Total energy (meV/atom) |
|---|---|---|
| A1 | Non-magnetic | 76.4 |
| | Paramagnetic | 0 |
| | Ferromagnetic | −51.3 |
| $L1_0$ | Non-magnetic | 109.8 |
| | Paramagnetic | −40.0 |
| | Ferromagnetic | −128.2 |

## 7.3 Electronic Structure

We now turn our attention to the multicomponent alloys of last chapter, comparing the results of the linear response analysis for the systems in different magnetic states. As before, the modelling starts with a *Hutsepot* calculation to model the electronic structure of the atomically disordered solid solution. We perform three sets of calculations. The first set are non-magnetic which, although expected to be unphysical for elements like Fe, are provided for reference. The second are the same as for the previous chapter, *i.e.*, modelling the paramagnetic state within the DLM picture. The third and final set of calculations are for a magnetically ordered state, ferrimagnetic for the alloys considered here.

Parameters for the self-consistent KKR-CPA calculations using the *Hutsepot* [14] code are set as in the previous chapters. Spin-polarised, scalar-relativistic calculations were performed within the atomic sphere approximation (ASA) [15], using an angular momentum cutoff of $l_{\text{max}} = 3$ for basis set expansions, a $20 \times 20 \times 20$ **k**-point mesh for integrals over the Brillouin zone, and a 24 point semi-circular Gauss-Legendre grid in the complex plane to integrate over valence energies. We use the local density approximation and the exchange-correlation functional is that of Perdew-Wang [16]. Lattice parameters for the systems were set at the experimental values [17–19] of 3.56, 3.57 and 3.59 Å, the same as for Chap. 6.

For these systems, within the DLM picture, Mn and Fe are found to support local moments, while Cr, Co, and Ni do not. When the magnetically ordered state is considered, for all three systems, the magnetic state predicted is *ferrimagnetic*; Cr and Mn always anti-align with the overall magnetic moment of the system, while Fe, Co, and Ni align with it. The sizes of the predicted magnetic moments in each system for the ferrimagnetic calculations are tabulated in Table 7.2 for reference. These elemental moments are in good agreement with the experimental measurements of [20].

In addition, for all systems studied in this chapter, it holds true that in our calculations

$$E_{\text{FM}} \leq E_{\text{DLM}} \leq E_{\text{non-magnetic}}, \tag{7.1}$$

**Table 7.2** Average magnetic moments associated with each chemical species in the magnetically ordered (but atomically *disordered*) solid solution for the considered multicomponent Cantor-Wu systems. Notably, in CrMnFeCoNi, Mn prefers to anti-align with the other elements present

| Material | Elemental Moment ($\mu_B$/atom) | | | | |
|---|---|---|---|---|---|
| | Cr | Mn | Fe | Co | Ni |
| CrCoNi | −0.163 | | | 0.832 | 0.148 |
| CrFeCoNi | −0.651 | | 1.922 | 1.079 | 0.282 |
| CrMnFeCoNi | −0.154 | −1.177 | 1.786 | 0.773 | 0.133 |

*i.e.*, a paramagnetic state is favoured over a non-magnetic state and, in turn, a ferrimagnetic state is energetically favourable compared to a paramagnetic state. The energy differences between the three considered magnetic states are tabulated in Table 7.3. We note that the energy differences between paramagnetic (PM) and ferrimagnetic (FM) states are small, reflective of the low, experimentally observed Curie temperatures of these systems.

In Fig. 7.1, we plot the density of states for CrCoNi, CrFeCoNi, CrMnFeCoNi for both non-magnetic, spin-polarised and DLM calculations. (Note that the DLM state for CrCoNi is, in fact, non-magnetic so the DoS is identical to that for the non-magnetic calculation.) It can be seen that the shape of the density of states curves around $E_F$ are significantly altered depending on the treatment of magnetism in the system.

Considering the three component CrCoNi first, where the DLM state collapses to a non-magnetic state, it can be seen that the ferrimagnetic calculation gives a markedly different DoS to the non-magnetic calculation. In the majority spin channel, the negative moment on Cr is manifested by its species-resolved curve moving to to higher energies, while the large positive moment on Co moves its species-resolved curve to lower energies and closer to that of Ni. These results are understood as resulting from the exchange splitting of the system's $d$-bands. It would not be unreasonable to suggest that, in the majority spin channel, the DoS begins to resemble that of $Ni_2Cr$, a system known to atomically order into an $MoPt_2$ crystal structure.

Proceeding, for the four-component CrFeCoNi, the energetic favourability of the DLM state over the non-magnetic state is demonstrated by the shift to lower energies of the species-resolved contribution associated with Fe, associated with local exchange splitting in the DLM state. Inspecting the results of the ferrimagnetic calculation, the species-resolved DoS contribution from Fe is significantly altered due to the large moment on Fe as a result of global exchange splitting, suggesting that the nature of interactions involving Fe will change when the chemical stability analysis is performed.

Finally, for CrMnFeCoNi, there are similar differences in the DoS for the non-magnetic calculation compared to the paramagnetic one as for CrFeCoNi. When considering the ferrimagnetic calculation, the anti-parallel aligned moment on Mn shifts its DoS curve in the majority spin channel to lie almost on top of that of Cr, while in the minority channel it lies almost on top of the averaged total DoS curve.

**Table 7.3** Calculated energy differences between the three considered magnetic states (Non-magnetic, DLM, FM) for the disordered solid solutions of each of CrCoNi, CrFeCoNi, and CrMnFeCoNi. For CrCoNi, the DLM state collapses to be non-magnetic

| Material | $E_{\mathrm{DLM}} - E_{\mathrm{Non\text{-}magentic}}$ (meV/atom) | $E_{\mathrm{FM}} - E_{\mathrm{DLM}}$ (meV/atom) |
|---|---|---|
| CrCoNi | 0 | −2.61 |
| CrFeCoNi | −11.78 | −14.24 |
| CrMnFeCoNi | −10.61 | −4.22 |

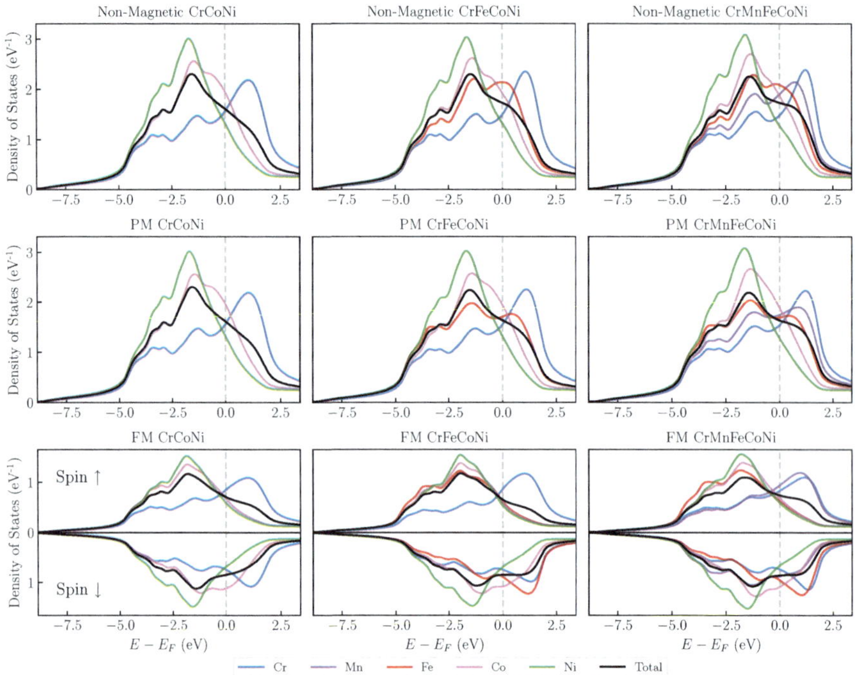


**Fig. 7.1** Density of states comparison for the Cantor-Wu alloys in non-magnetic (top row), paramagnetic (middle row), and ferrimagnetic (bottom row) states. The paramagnetic state is modelled within the DLM picture

These results are again understood to originate in local/global exchange splitting, and we anticipate these changes in the underlying electronic structure of the material to alter predicted atomic short-range order.

## 7.4 Linear Response

We use the self-consistent potentials for the three different treatments of magnetism to calculate the $S^{(2)}_{\alpha\alpha'}(\mathbf{k})$ and, subsequently, to construct the chemical stability matrix in reciprocal space, before calculating its eigenvalues, as in previous chapters. These results of which are visualised in Figs. 7.2 and 7.3.

Considering the $S^{(2)}_{\alpha\alpha'}(\mathbf{k})$ data first (Fig. 7.2) we see that the nature of atom-atom correlations is substantially altered depending on the treatment of magnetism. This effect is most pronounced for those correlations involving Fe and Mn, which makes sense because these are the elements which typically support local moments in these fcc solid solutions. We see that the $S^{(2)}$ data associated with Cr-Fe, Fe-Ni, Cr-Mn,

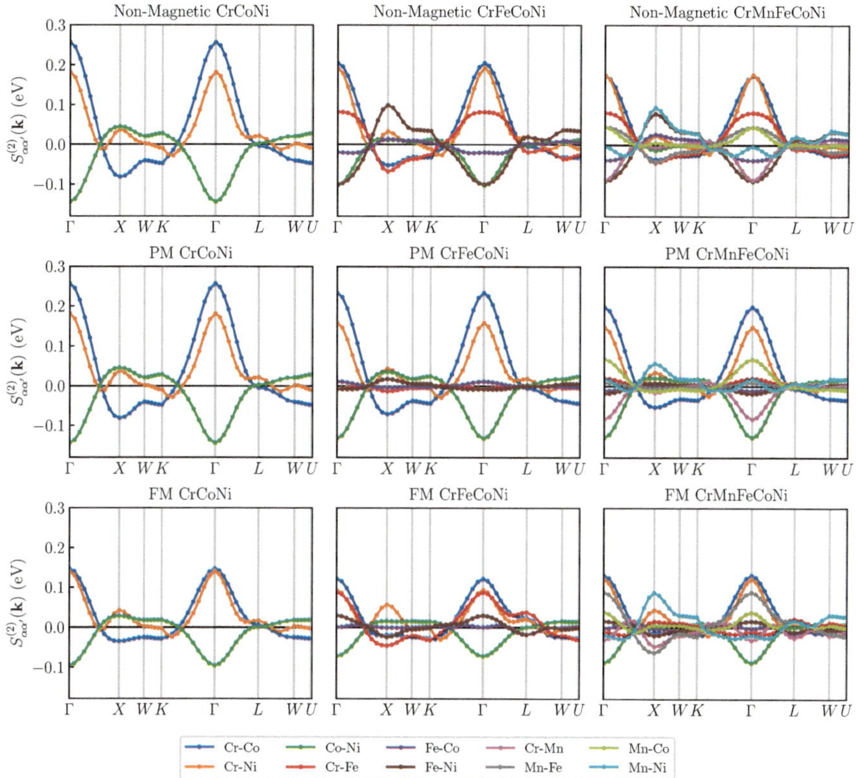


**Fig. 7.2** Comparison of $S^{(2)}_{\alpha\alpha'}(\mathbf{k})$ around the IBZ of the fcc lattice for linear-response calculations performed on non-magnetic (top row), paramagnetic (middle row), and ferrimagnetic (bottom row) potentials. It can be seen that, particularly for elements such as Fe and Mn, there is a substantial shift in the character of $S^{(2)}_{\alpha\alpha'}(\mathbf{k})$ depending on the magnetic state

and Mn-Ni pairs is substantially altered in character depending on the treatment of magnetism, which will impact predicted atomic ordering.

Proceeding, we move on to consider the eigenvalues of the chemical stability matrices, visualised in Fig. 7.3. In the previous chapter that employed only paramagnetic calculations, we portrayed a simple picture of Cr-Co interactions as dominant, while Mn and Fe interact only weakly, contributing near-flat concentration wave modes and diluting the strength of interactions.[1] For the non-magnetic calculations, this picture can still be seen to hold to some extent, but the picture for the systems modelled assuming a magnetically ordered state differs significantly from this; the flat concentration wave mode associated with Fe in CrFeCoNi is completely changed in character, and in CrMnFeCoNi there is only one flat mode now, not two. This lifting

[1] Recall that a near-flat concentration wave mode is associated with weak interactions and, therefore, with weak ASRO.

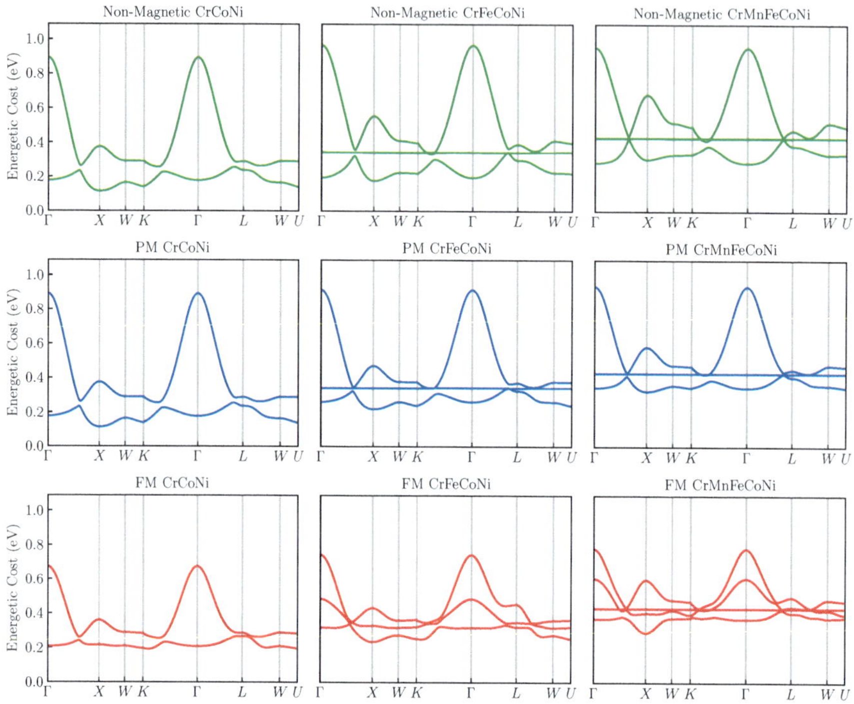


**Fig. 7.3** Comparison of the eigenvalues of the chemical stability matrix for the considered Cantor-Wu alloys based on linear response calculations performed on non-magnetic (top row), paramagnetic (middle row), and ferrimagnetic (bottom row) potentials

of degeneracy of the two flat modes is associated with stronger atomic correlations between Fe, Mn and other elements present.

In Fig. 7.3, for the ternary CrCoNi alloy, clear competition can be seen between energetic minima at $\mathbf{k} = (0, 0, 1)$, indicative of $L1_2$-like order, and $\mathbf{k} = (0, \frac{2}{3}, \frac{2}{3})$, associated with $MoPt_2$-like order. When modelling the magnetically ordered state, the minimum at $(0, \frac{2}{3}, \frac{2}{3})$ sits lower than that at $(0, 0, 1)$, but for the magnetically disordered[2] state it is the minimum at $(0, 0, 1)$ which is at lower energy and therefore describes the dominant ASRO. This outcome highlights the competition between $L1_2$ and $MoPt_2$ order in this system, as has been noted in earlier works [7, 21], but also emphasises the connection between this competition and magnetism, as discussed by Walsh *et al.* [4].

Predicted compositional ordering temperatures, concentration wave modes, and chemical polarisations for the alloys, assuming either a non-magnetic, paramagnetic, or ferrimagnetic state, are tabulated in Table 7.4. For clarity, we visualise the different partially ordered structures for these alloys in Fig. 7.4. Structures with partial atomic ordering are obtained by allowing the concentration wave predicted by the chemical

[2] Recall that, for this alloy, the paramagnetic state and non-magnetic state are equivalent.

**Table 7.4** Transition temperatures, modes, and chemical polarisations for the considered Cantor-Wu systems generated from paramagnetic (DLM) potentials, non-magnetic potentials, and ferrimagnetic potentials. There are differences in the temperature and nature of predicted compositional order between the magnetic states. Notably, for CrCoNi, the mode shifts from $\mathbf{k} = (0, 0, 1)$ to $\mathbf{k} = (0, \frac{2}{3}, \frac{2}{3})$, a point along the line from $\Gamma$ to $K$, indicative of a shift from $L1_2$-like to $MoPt_2$-like order. The small numerical differences between these values and [22, 23] are due to alterations to the numerical solver

| Material | Magnetic state | $T_{us}(K)$ | $\mathbf{k}_{us}$ | $\Delta c_{Cr}$ | $\Delta c_{Mn}$ | $\Delta c_{Fe}$ | $\Delta c_{Co}$ | $\Delta c_{Ni}$ |
|---|---|---|---|---|---|---|---|---|
| CrCoNi | Non-magnetic | 581 | (0, 0, 1) | 0.723 | | | −0.690 | −0.032 |
| | Paramagnetic | 581 | (0, 0, 1) | 0.723 | | | −0.690 | −0.032 |
| | Ferrimagnetic | 252 | (0, 2/3, 2/3) | 0.813 | | | −0.468 | −0.345 |
| CrFeCoNi | Non-magnetic | 399 | (0, 0, 0) | 0.080 | | −0.613 | −0.221 | 0.754 |
| | Paramagnetic | 380 | (0, 0, 1) | 0.722 | | −0.049 | −0.686 | 0.016 |
| | Ferrimagnetic | 313 | (0, 0, 1) | 0.531 | | −0.787 | −0.052 | 0.308 |
| CrMnFeCoNi | Non-magnetic | 299 | (0, 0, 0) | 0.220 | −0.376 | −0.467 | −0.135 | 0.758 |
| | Paramagnetic | 258 | (0, 0, 1) | 0.722 | 0.010 | −0.082 | −0.686 | 0.035 |
| | Ferrimagnetic | 319 | (0, 0, 1) | 0.227 | −0.625 | 0.706 | −0.077 | −0.231 |

instability analysis to 'grow' until one sublattice contains (at least) one atomic species whose concentration reaches zero. This condition identifies the the largest permitted chemical fluctuation consistent with that polarisation. Note that, as the non-magnetic and DLM states for CrCoNi are equivalent, we only visualise this structure once.

Considering first the non-magnetic results for CrFeCoNi and CrMnFeCoNi, it can be seen that the minimum eigenvalue at $\Gamma$ is found to pass through 0 first, indicative of phase segregation. The chemical polarisation suggests that in CrFeCoNi, Ni will tend to segregate away from Fe and Co, whilst in CrMnFeCoNi, Ni and Cr will segregate away from Mn, Fe, and Co. We suggest that the picture will be more subtle than this naive description, however, as the minimum eigenvalue at the $X$ point lies very close in energy to the minimum at $\Gamma$, indicative of competition between modes.

Considering then the differences in predicted order between paramagnetic and ferrimagnetic calculations, the most striking change in predicted atomic ordering is noted for the ternary CrCoNi alloy, where the calculated atomic disorder-order temperature for a magnetically ordered state is significantly lower than that for the same system with magnetic disorder. In this case the predicted structure is now $MoPt_2$-like rather than $L1_2$-like. In the four-component CrFeCoNi alloy, while the wave-vector describing the order remains the same, magnetic order is found to change the chemical polarisation of the mode significantly, strengthening correlations involving Fe atoms and weakening those involving Co atoms. These results are now in good agreement with the study by Tamm *et al.* [2], where Monte Carlo simulations with energies evaluated via spin-polarised DFT calculations found strong correlations between Fe and other elements, a result which we failed to find when modelling the paramagnetic state. This spin-driven ordering, with Cr atoms moving to occupy the corners of the fcc lattice, is also in alignment with the study of Niu *et al.* [5]. There is also agreement

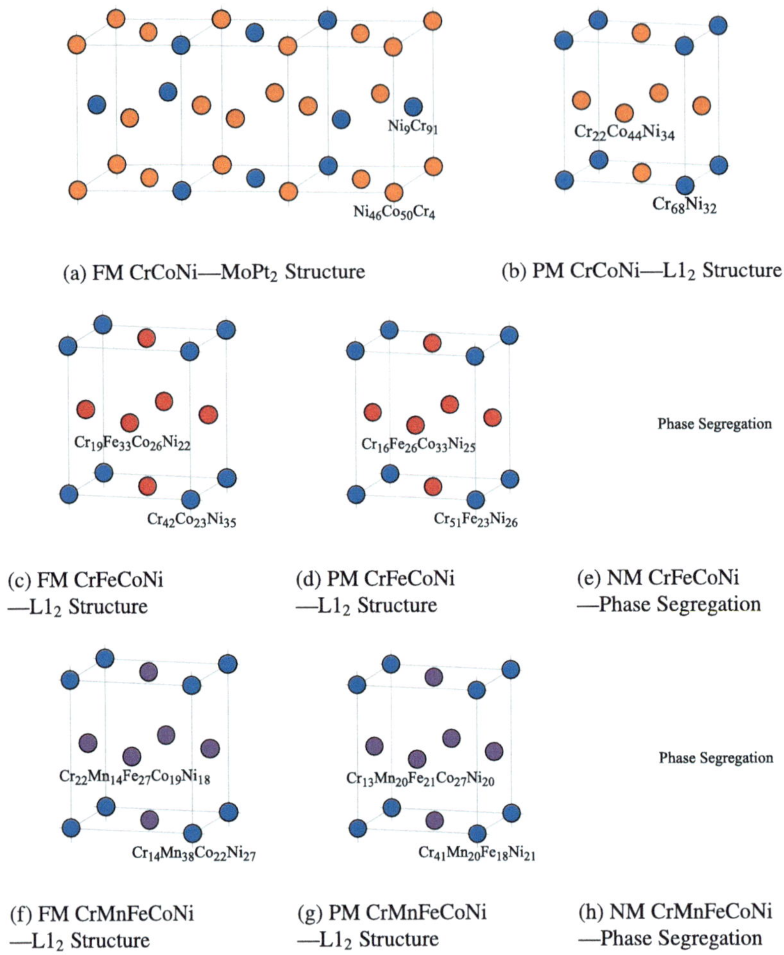


**Fig. 7.4** Comparison of the partially ordered structures predicted with our linear response theory for CrCoNi, CrFeCoNi, and CrMnFeCoNi in their magnetically ordered (ferrimagnetic, 'FM') magnetically disordered (paramagnetic, 'PM') and non-magnetic ('NM') states. For CrCoNi, both orderings are driven by Cr, but in the magnetically ordered state, $MoPt_2$-like order is predicted, while in the paramagnetic state it is $L1_2$-like order that is favoured. For CrFeCoNi and CrMnFeCoNi, we see that the nature of the $L1_2$ ordering is altered in character depending on whether the system is treated as paramagnetic or ferrimagnetic. Note that we do not show non-magnetic CrCoNi; this and the paramagnetic state are equivalent as none of the elements support a local moment in this composition

with a study by Meshkov *et al.* [24], who use a machine-learned interatomic potential trained on spin-polarised DFT calculations to find that both Cr and Fe atoms set up on distinct sublattices in the CrFeCoNi alloy, a result broadly in alignment with ours.

However, these authors predicts an initial atomic ordering temperature of 1230°C (= 1503 K), which is far higher than ours.

A similar story to the CrFeCoNi system holds true in the complex CrMnFeCoNi alloy, where correlations involving both Fe and Mn, which both acquire large magnetic moments, are also strengthened compared to the results obtained in our earlier work modelling the paramagnetic state [22]. We find that this $L1_2$ ordering is now driven by Mn, which is understood by considering that the magnetic moment of Mn, similarly to Cr, aligns anti-parallel with the total moment of the system, but with a much larger moment. Placing Mn on the corners of the fcc lattice allows each Mn atom (moment anti-parallel) to be surrounded primarily by elements whose moment align parallel with the total moment of the system. This ordering is, therefore, clearly spin-driven. This result is broadly in alignment with the study of Mizuno *et al.* [25], which predicts an $L1_2$ ordering with Cr and Mn atoms moving to the corner sites of the fcc lattice in the $L1_2$ structure.

One notable disagreement between our results and those of earlier works is described in the recent study by Ghosh *et al.* [7], which predicted $MoPt_2$-like order emerging at 975 K in CrCoNi. Simulations using low-rank potentials[3] have previously been in good agreement with results obtained using our approach when modelling refractory HEAs [26, 27], so this discrepancy is surprising. We note that a recent experimental study by Li *et al.* [28] performed annealing of CrCoNi at a variety of temperatures, including one sample annealed at 673 K for 500 h. They found evidence of increasing ASRO in their samples, but no evidence of a long-range ordered state was reported. Further, the known $MoPt_2$-former, $CrNi_2$, has an experimentally observed atomic ordering temperature of 863 K, which is in good agreement with the theoretical prediction around 880 K [29, 30]. We would expect the atomic ordering temperature for the ternary CrCoNi alloy to be lower than that predicted for $CrNi_2$, because the increased entropy of mixing of the ternary alloy over that of the binary alloy will stabilise the solid solution.

## 7.5 Pairwise Interactions

To understand further the nature of atomic short-range order present below any initial ordering temperature for each magnetic state, we once again fit a real-space interaction to our reciprocal space $S^{(2)}_{\alpha\alpha'}(\mathbf{k})$ data to recover a model suitable for lattice-based atomistic modelling. We sample $S^{(2)}_{\alpha\alpha'}(\mathbf{k})$ at 56 **k**-points around the irreducible Brillouin zone, including high-symmetry points.

Visualised in Fig. 7.5 are plots of $V^{(n)}_{\alpha\alpha'}$ as a function of distance for the first ten coordination shells for each considered magnetic state. While it can be seen that the *range* of interactions is largely unaffected—interactions are always dominated by the first four coordination shells—the *nature* of interactions is altered. This is evidenced

[3] Low-rank potentials (LRPs) are a class of machine-learned interatomic potentials [26].

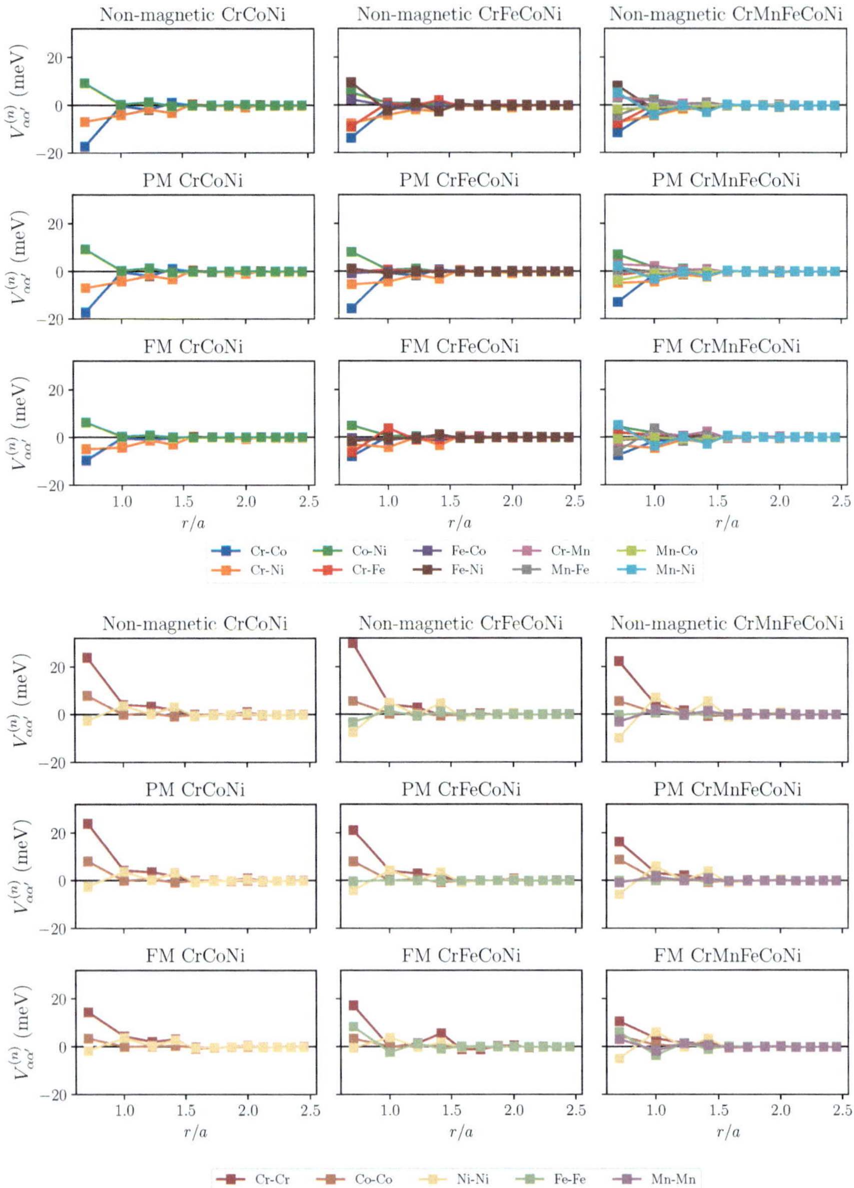


**Fig. 7.5** Visualisation of $V_{\alpha\alpha'}^{(n)}$ for the three multicomponent Cantor-Wu systems considered, fitted to the first ten coordination shells. It can be seen that interactions are dominated by the first four coordination shells

by significant changes in pairwise interactions, particularly between Fe, Cr, and the other elements present.

Based on the above observations, we fit to $V_{i\alpha;j\alpha'}$ to the first four coordination shells of the fcc lattice (consistent with Chap. 6). The fitted atom-atom interactions are tabulated in Appendix C, Tables C.10–C.17.

## 7.6 Atomistic Modelling

Using the fitted $V_{i\alpha;j\alpha'}$, we perform lattice-based Monte Carlo simulations as outlined in Chap. 2 for each of the three systems in each of the three magnetic states. These simulations were all performed using a system consisting of $12 \times 12 \times 12$ cubic fcc unit cells, each with 4 lattice sites per unit cell, for a total of 6192 atoms. Periodic boundary conditions were applied. The systems were prepared in an initially random configuration, then computationally annealed from 1200 K to 10 K in steps of 1 K, with $10^4$ MC steps per atom at each temperature. For each chemical composition and magnetic state, an ensemble of 10 simulations was performed and the results averaged, to extract uncertainty estimates on key quantities and to improve the statistics.

Figure 7.6 compares the Warren-Cowley ASRO parameters on first and second coordination shells, *i.e.*, for first- and second-nearest neighbours. Results are shown for interactions computed based on the non-magnetic state, the magnetically disordered (paramagnetic) state, and for those computed based on the magnetically ordered (ferrimagnetic) state. There are clear differences in the strength and nature of predicted ASRO in all three alloy compositions, depending on which magnetic state is simulated. Although still favoured, the Cr-Co correlations in CrCoNi are significantly weakened when the system is modelled under the assumption of magnetic order. The favouring of Cr-Co pairs on the first neighbour shell is consistent with the results of earlier computational studies [2–4, 7, 21].

In CrFeCoNi, when the non-magnetic state is simulated, the dominant correlations are Fe-Ni, Co-Ni, and Fe-Co. This is in contrast to the results for the paramagnetic state, where Fe-Ni and Fe-Co correlations are significantly weakened, and instead it is Cr-Co and Co-Ni correlations which dominate. Proceeding, for the ferrimagnetic state, Cr-Co correlations are also weakened, while correlations between Fe and all three other elements are significantly strengthened. This outcome now agrees qualitatively with the work by Tamm *et al.* [2] as well as with other previous works [5, 6, 31].

Finally, while the picture is less clear-cut for CrMnFeCoNi, it can be seen that there are significant qualitative differences in ASRO between all three magnetic states. In particular, we note that Mn-Fe and Mn-Ni atomic correlations dominant in the low temperature regime when the system is modelled in a magnetically ordered state, which is in contrast the results of Chap. 6 for the paramagnetic state. We again note that there is limited information in the literature to allow assessment of our results obtained for the five-component alloy system, as the huge potential configuration

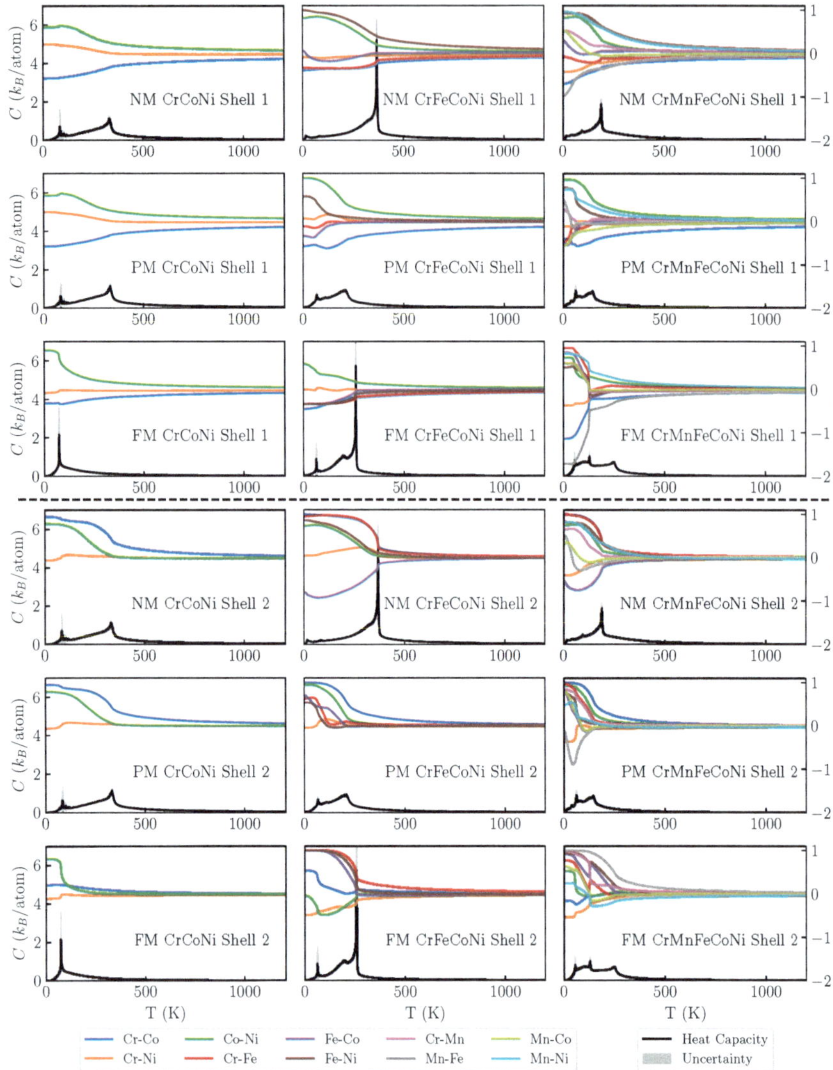


**Fig. 7.6** Warren-Cowley ASRO comparison for the Cantor-Wu alloys. The top 3 × 3 block shows results on the first shell, while the bottom block shows results for the second coordination shell. In each block, the top row shows non-magnetic results, the middle row paramagnetic, and the bottom row ferrimagnetic

space challenges most conventional supercell-based computational techniques. The two studies by Mizuno *et al.* [25], using DFT-based Monte Carlo simulations, and Kar'kin *et al.* [32], using the modified embedded atom method, are in agreement with our result that correlations between Cr and Mn (magnetic moment anti-parallel) and

Fe, Co, and Ni (magnetic moment parallel) are strongest, once again suggesting a spin-driven ordering.

These results have clear implications for the materials modelling community as well as for experimental researchers. It is confirmed that due consideration must be given to the magnetic state of a material when modelling multicomponent alloy systems containing 3*d* transition metals. The presence and nature of magnetic order significantly impacts the strength and nature of predicted ASRO. It may also be true that there are some high-entropy alloy compositions for which the application of a magnetic field during the annealing process, especially at temperatures below any magnetic transition temperature(s), will impact the ASRO by affecting the material's magnetic state and in turn impact its mechanical properties.

## 7.7 Summary

In summary, results have been presented that shed light on the importance of the correct treatment of a system's magnetic state when modelling multicomponent alloys containing 3*d* transition metals. Depending on how the magnetic state is computationally treated, significant differences in predictions of short- and long-range order are noted for all three multicomponent alloy systems considered in this study. In particular, our results reiterate that competition between $L1_2$-like and $MoPt_2$-like order in CrCoNi is unequivocally connected to magnetism in the system, as has been suggested in earlier works. In addition, we note that our findings that Fe and Mn interact strongly when treated in a magnetically ordered (ferrimagnetic) state, is in alignment with other studies suggesting magnetism drives ordering in the CrFeCoNi and CrMnFeCoNi alloys.

These results suggest that studies employing supercell calculations should take care to treat the magnetic state of a system correctly when considering energy/force calculations. This may be particularly relevant when training machine-learned interatomic potentials aiming to study both the low- and high-temperature phase behaviour and dynamics of these systems, because a paramagnetic state may be a more accurate description of the high-temperature regime, while a magnetically ordered state may be a more accurate description of the system at lower temperatures. We suggest, therefore, that machine-learned interatomic potentials for alloys containing magnetic elements, such as those considered in this chapter, should aim to incorporate magnetism and magnetic degrees of freedom in an appropriate manner to ensure physically meaningful, and reliable, results are obtained. Examples of schemes aiming in incorporate these effects are Magnetic Moment Tensor Potentials [33] and Spin Gaussian Approximation Potentials [34].

In addition, our outlined technique for computational modelling of magnetic alloys in both magnetically ordered and magnetically disordered states has potential to inform experiment by suggesting whether multicomponent alloys with sufficiently high Curie temperatures should be annealed above or below $T_C$, or in an applied magnetic field, to promote desired ordering and/or other beneficial materials properties.

Our results therefore have implications both for materials modelling as well as for materials preparation. An interesting avenue of further study on the CrCoNi system would be to model how magnetic interactions in this system vary in different atomically ordered structures when compared to the disordered solid solution to further elucidate the connection between magnetism and ASRO. Another avenue of work to pursue experimentally would be to anneal some medium- and high-entropy alloy compositions in an applied magnetic field to assess the degree to which ASRO, and therefore materials properties, can be tuned.

## References


1. E. Mendive-Tapia, J. Neugebauer, T. Hickel, Phys. Rev. B **105**(6), 064425 (2022). https://doi.org/10.1103/PhysRevB.105.064425
2. A. Tamm, A. Aabloo, M. Klintenberg, M. Stocks, A. Caro, Acta Materialia **99**, 307 (2015). https://doi.org/10.1016/j.actamat.2015.08.015
3. J. Ding, Q. Yu, M. Asta, R.O. Ritchie, Proceed. Nat. Acad. Sci. **115**(36), 8919 (2018). https://doi.org/10.1073/pnas.1808660115
4. F. Walsh, M. Asta, R.O. Ritchie, Proceed. Nat. Acad. Sci. **118**(13), e2020540118 (2021). https://doi.org/10.1073/pnas.2020540118
5. C. Niu, A.J. Zaddach, A.A. Oni, X. Sang, J.W. Hurt, J.M. LeBeau, C.C. Koch, D.L. Irving, Appl. Phys. Lett. **106**(16), 161906 (2015). https://doi.org/10.1063/1.4918996
6. T. Fukushima, H. Katayama-Yoshida, K. Sato, M. Ogura, R. Zeller, P.H. Dederichs, J. Phys. Soc. Japan **86**(11), 114704 (2017). https://doi.org/10.7566/JPSJ.86.114704
7. S. Ghosh, V. Sotskov, A.V. Shapeev, J. Neugebauer, F. Körmann, Phys. Rev. Mater. **6**(11), 113804 (2022). https://doi.org/10.1103/PhysRevMaterials.6.113804
8. K.E. Guthe, Trans. American Inst. Electr. Eng. **XIV**(1), 55 (1897). https://doi.org/10.1109/T-AIEE.1897.5570176
9. H. Pender, R.L. Jones, Phys. Rev. **1**(4), 259 (1913). https://doi.org/10.1103/PhysRev.1.259
10. R. McCurrie, in *Handbook of Ferromagnetic Materials*, vol. 3 (Elsevier, 1982), pp. 107–188. https://linkinghub.elsevier.com/retrieve/pii/S1574930405800896
11. K. Suzuki, N. Kataoka, A. Inoue, A. Makino, T. Masumoto, Mater. Trans. JIM **31**(8), 743 (1990). https://doi.org/10.2320/matertrans1989.31.743
12. Y. Yoshizawa, S. Oguma, K. Yamauchi, J. Appl. Phys. **64**(10), 6044 (1988). https://doi.org/10.1063/1.342149
13. L. Néel, J. Pauleve, R. Pauthenet, J. Laugier, D. Dautreppe, J. Appl. Phys. **35**(3), 873 (1964). https://doi.org/10.1063/1.1713516
14. M. Hoffmann, A. Ernst, W. Hergert, V.N. Antonov, W.A. Adeagbo, R.M. Geilhufe, H. Ben Hamed, Physica Status Solidi (b) **257**(7), 1900671 (2020). https://doi.org/10.1002/pssb.201900671
15. G.M. Stocks, W.M. Temmerman, B.L. Gyorffy, Phys. Rev. Lett. **41**(5), 339 (1978). https://doi.org/10.1103/PhysRevLett.41.339
16. J.P. Perdew, Y. Wang, Phys. Rev. B **45**(23), 13244 (1992). https://doi.org/10.1103/PhysRevB.45.13244
17. Y. Zhang, G.M. Stocks, K. Jin, C. Lu, H. Bei, B.C. Sales, L. Wang, L.K. Béland, R.E. Stoller, G.D. Samolyuk, M. Caro, A. Caro, W.J. Weber, Nat. Commun. **6**(1), 8736 (2015). https://doi.org/10.1038/ncomms9736

18. B. Yin, S. Yoshida, N. Tsuji, W.A. Curtin, Nat. Commun. **11**(1), 2507 (2020). https://doi.org/10.1038/s41467-020-16083-1
19. B. Cantor, I.T.H. Chang, P. Knight, A.J.B. Vincent, Mater. Sci. Eng. A **375–377**, 213 (2004). https://doi.org/10.1016/j.msea.2003.10.257
20. D. Billington, A.D.N. James, E.I. Harris-Lee, D.A. Lagos, D. O'Neill, N. Tsuda, K. Toyoki, Y. Kotani, T. Nakamura, H. Bei, S. Mu, G.D. Samolyuk, G.M. Stocks, J.A. Duffy, J.W. Taylor, S.R. Giblin, S.B. Dugdale, Phys. Rev. B **102**(17), 174405 (2020). https://doi.org/10.1103/PhysRevB.102.174405
21. J.P. Du, P. Yu, S. Shinzato, F.S. Meng, Y. Sato, Y. Li, Y. Fan, S. Ogata, Acta Materialia **240**, 118314 (2022). https://doi.org/10.1016/j.actamat.2022.118314
22. C.D. Woodgate, J.B. Staunton, Phys. Rev. B **105**(11), 115124 (2022). https://doi.org/10.1103/PhysRevB.105.115124
23. C.D. Woodgate, D. Hedlund, L.H. Lewis, J.B. Staunton, Phys. Rev. Mater. **7**(5), 053801 (2023). https://doi.org/10.1103/PhysRevMaterials.7.053801
24. E. Meshkov, I. Novoselov, A. Shapeev, A. Yanilkin, Intermetallics **112**, 106542 (2019). https://doi.org/10.1016/j.intermet.2019.106542
25. M. Mizuno, K. Sugita, H. Araki, Res. Phys. **34**, 105285 (2022). https://doi.org/10.1016/j.rinp.2022.105285
26. T. Kostiuchenko, F. Körmann, J. Neugebauer, A. Shapeev, Npj Comput. Mater. **5**(1), 55 (2019). https://doi.org/10.1038/s41524-019-0195-y
27. C.D. Woodgate, J.B. Staunton, Phys. Rev. Mater. **7**(1), 013801 (2023). https://doi.org/10.1103/PhysRevMaterials.7.013801
28. L. Li, Z. Chen, S. Kuroiwa, M. Ito, K. Yuge, K. Kishida, H. Tanimoto, Y. Yu, H. Inui, E.P. George, Acta Mater. **243**, 118537 (2023). https://doi.org/10.1016/j.actamat.2022.118537
29. J.B. Staunton, D.D. Johnson, F.J. Pinski, Phys. Rev. B **50**(3), 1450 (1994). https://doi.org/10.1103/PhysRevB.50.1450
30. M. Rahaman, B. Johansson, A.V. Ruban, Phys. Rev. B **89**(6), 064103 (2014). https://doi.org/10.1103/PhysRevB.89.064103
31. B. Schönfeld, C.R. Sax, J. Zemp, M. Engelke, P. Boesecke, T. Kresse, T. Boll, T. Al-Kassab, O.E. Peil, A.V. Ruban, Phys. Rev. B **99**(1), 014206 (2019). https://doi.org/10.1103/PhysRevB.99.014206
32. I.N. Kar'kin, L.E. Kar'kina, Y.N. Gornostyrev, Phys. Metals Metallogr. **124**(10), 1026 (2023). https://doi.org/10.1134/S0031918X23601671
33. I. Novikov, B. Grabowski, F. Körmann, A. Shapeev, Npj Comput. Mater. **8**(1), 13 (2022). https://doi.org/10.1038/s41524-022-00696-9
34. L. Shenoy, C.D. Woodgate, J.B. Staunton, A.P. Bartók, C.S. Becquart, C. Domain, J.R. Kermode, Phys. Rev. Mater. **8**(3), 033804 (2024). https://doi.org/10.1103/PhysRevMaterials.8.033804

# Chapter 8
# Compositional Order and Subsequent Magnetostriction in $Fe_{1-x}Ga_x$ (Galfenol)

**Abstract** As an example of how atomic arrangements in an alloy can affect materials properties, we consider the effect of atomic ordering in the binary $Fe_xGa_{1-x}$ alloy, commonly referred to as Galfenol. We show that the ordered structures of interest are the B2 and $D0_3$ structures, and examine the influence of these chemical orderings on the Curie temperature and magnetoelastic constant of this system within a density functional theory (DFT) framework. We are able to show that these chemical orderings reduce the predicted Curie temperature of the system, but bring predicted magnetoelastic constants in to line with the measured experimental values. These results serve as an example of how atomic ordering can affect the magnetic properties of an alloy.

## 8.1 Introduction

The binary alloy $Fe_{1-x}Ga_x$, commonly referred to as Galfenol, is of interest to the materials science community because it exhibits high levels of magnetostriction [1–3]. Magnetostriction is a phenomenon where a material deforms under the application of an external magnetic field [4]. It finds application in the development of sensor and actuator technologies because it provides a mechanism by which to convert energy from mechanical to electrical and vice-versa. Pure iron is already magnetostrictive, but experimentally it is found that by adding a small amount of Ga to the material ($x \sim 0.2$), this property is enhanced, and measured quantities of interest are around an order of magnitude greater [5]. Moreover, theoretical work has demonstrated that the assumption that the added Ga is spread randomly across the lattice is flawed and fails to explain the enhancement of the magnetostriction [6]. A complete description of atomic arrangements of this system is therefore needed.

The supercell approach allows one to study the properties of a particular ordered system, and has been applied to the Galfenol system with some success [7, 8]. Indeed, one recent work made use of *ab initio* molecular dynamics calculations to obtain ordered structures, the magnetic properties of which they were then able to study [9]. These approaches study particular configurations of the system with a specified atom on each site. However, due to the constraint on the number of atoms



C. D. Woodgate, *Modelling Atomic Arrangements in Multicomponent Alloys*,
Springer Series in Materials Science 346,
https://doi.org/10.1007/978-3-031-62021-8_8

in the system these techniques can only sample discrete concentrations, and it is also difficult to be sure whether the configurations studied are truly reflective of the ensemble average of the system configuration.

In this chapter, we use our approach for modelling atomic arrangements in multi-component alloys to study the compositional phase behaviour of this material in both its paramagnetic and ferromagnetic states. We also use it as an example application of the 'lattice gas' model. Having obtained a description of atomic arrangements in the material, we then go on to examine the impact of these atomic arrangements on the system's magnetic properties. We provide results for the Curie temperature of the material and show that the Curie temperature of ordered phases is reduced compared to the disordered phase. In addition, we calculate the system's magnetoelastic coefficient, and successfully reproduce a hitherto unexplained peak in its value at $x \sim 0.19$ [10].

## 8.2 A Primer on Magnetostriction

A magnetostrictive material is one which deforms on the application of a magnetic field. This can be described by the competing effects of the (magnetisation independent) elastic energy, $E_{\mathrm{el}}$, and the magnetoelastic energy, $E_{\mathrm{mag}}$ [11]. In a cubic material under strain given by strain tensor $\varepsilon_{ij}$, the elastic energy takes the form

$$E_{\mathrm{el}}=\frac{1}{2}c_{11}(\varepsilon_{xx}^2+\varepsilon_{yy}^2+\varepsilon_{zz}^2)+\frac{1}{4}c_{44}(\varepsilon_{xy}^2+\varepsilon_{yz}^2+\varepsilon_{zx}^2)+c_{12}(\varepsilon_{xx}\varepsilon_{yy}+\varepsilon_{yy}\varepsilon_{zz}+\varepsilon_{xx}\varepsilon_{zz}), \tag{8.1}$$

where $c_{ij}$ are elastic constants to be determined. The magnetoelastic energy is then given by

$$E_{\mathrm{mag}} = B_1(\alpha_x^2\varepsilon_{xx}+\alpha_y^2\varepsilon_{yy}+\alpha_z^2\varepsilon_{zz}) + B_2(\alpha_x\alpha_y\varepsilon_{xy}+\alpha_y\alpha_z\varepsilon_{yz}+\alpha_x\alpha_z\varepsilon_{xz}), \tag{8.2}$$

where $\{\alpha_i\}$ are the direction cosines of the magnetisation direction, $\mathbf{n} = (\alpha_x, \alpha_y, \alpha_z)$, and $B_1$ and $B_2$ are associated magnetoelastic constants. By balancing these two separate contributions to the total energy, it is then possible to derive expressions for the dimensionless cubic magnetostriction constants [6],

$$\lambda_{001} = -\frac{2}{3}\frac{B_1}{c_{11}-c_{12}} \tag{8.3}$$

and

$$\lambda_{111} = -\frac{1}{3}\frac{B_2}{c_{44}}. \tag{8.4}$$

It should be noted that these results are valid under the assumption that the material's elastic constants are independent of magnetisation. For Galfenol, the more relevant of the two magnetostriction constants is known to be the tetragonal constant, $\lambda_{001}$ [12].

For pure Fe, $\lambda_{001}$ is approximately $30 \times 10^{-6}$, while when alloyed with Ga at a concentration of 17%, it is almost an order of magnitude greater, at $207 \times 10^{-6}$ [12].

To determine $B_1$ within an *ab initio* computational framework, we begin by considering the, magnetic torque, which has its origins in a system's magnetocrystalline anisotropy [6, 10]. We write the torque as

$$\mathbf{T}^{(\hat{\mathbf{n}})} = -\frac{\partial \mathcal{E}^{(\hat{\mathbf{n}})}}{\partial \hat{\mathbf{n}}}, \tag{8.5}$$

where $\hat{\mathbf{n}}$ denotes the magnetisation direction and $\mathcal{E}$ the relativistic DFT energy of the system. This quantity can be obtained within a relativistic DFT framework, which is discussed in Appendix B. For a system strained along the $\hat{\mathbf{z}}$ direction with strain $\varepsilon_{zz}$, it can be shown that evaluation of the torque for a polar angle of $\theta = \pi/4$ yields the anisotropy constant $B_1$ multiplied by the strain. It is therefore possible to apply a range of strains and fit to a value of $B_1$ for a given system. A more detailed discussion of the procedure for obtaining the magnetoelastic constants can be found in [6, 10]. Crucially, the technique is applicable to a variety of crystal structures, making it possible to assess the impact of chemical ordering on a material's magnetoelasticity.

## 8.3 Electronic Structure

As ever, the first stage of the modelling procedure is to obtain a description of the material's electronic structure. For this system, we wish to consider a range of concentrations of Ga. Because Ga is a larger atom than Fe, higher concentrations of Ga result in a marginally expanded lattice [13]. We note that, for the range of Ga concentrations considered in this chapter, the underlying lattice is always bcc [13].

Taking the data of [13, 14], we fit a simple quadratic to the data for the lattice expansion, which captures the experimental data well. The data and fit are visualised in Fig. 8.1. Our fitted lattice parameters, used in subsequent calculations, are tabulated

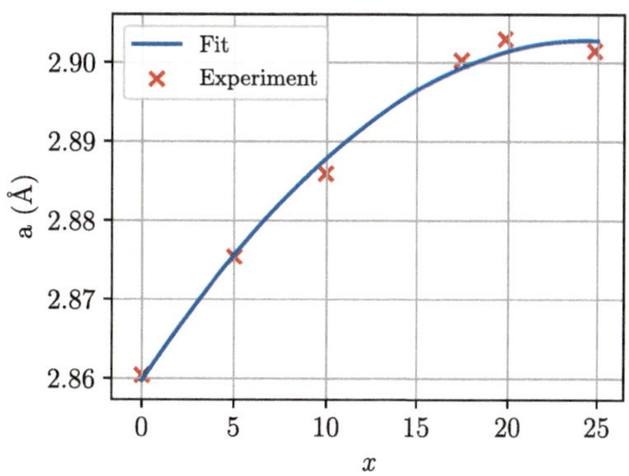


**Fig. 8.1** Experimental data for the lattice spacing of $Fe_{1-x}Ga_x$ as a function of $x$ (red crosses) compared with the fitted values (blue line). The expansion of the lattice is due to the larger relative size of Ga compared to Fe

**Table 8.1** Fitted lattice parameters for the $Fe_{1-x}Ga_x$ system as visualised in Fig. 8.1

| $x$ | a (Å) |
|---|---|
| 0 | 2.860 |
| 0.05 | 2.876 |
| 0.1 | 2.888 |
| 0.15 | 2.896 |
| 0.2 | 2.901 |
| 0.25 | 2.903 |

in Table 8.1. We note that there are no quoted experimental errors for the measured lattice parameters. It could be that, given suitable error bars, a linear fit could also be appropriate.

Once again, we then use the *Hutsepot* code [15] to obtain the one-electron self-consistent potentials of DFT. The parameters and settings for the self-consistent calculations are the same as for the calculations of previous chapters. Spin-polarised, scalar-relativistic calculations were performed within the atomic sphere approximation (ASA) [16], using an angular momentum cutoff of $l_{\max} = 3$ for basis set expansions, a $20 \times 20 \times 20$ **k**-point mesh for integrals over the Brillouin zone, and a 24 point semi-circular Gauss-Legendre grid in the complex plane to integrate over valence energies. We use the local density approximation and the exchange-correlation functional is that of Perdew-Wang [17].

Visualised in Fig. 8.2 are the total and species-resolved density of states around $E_F$ for $x = 0, 0.1, 0.2$ in both paramagnetic (DLM) and ferromagnetic states. The electronic structure of Galfenol is different to the systems we have considered in earlier chapters. The 3*d* states associated with Ga are all filled and should be thought of as localised core states, while its valence electrons are best thought of as *s*- and *p*-like. However, the filled 3*d* shell is very close in energy to the valence band. When alloyed with Fe, whose valence electrons *are* *d*-like, the gap in energy between the 3*d* states of Ga and the valence band is reduced further, and the 3*d* states of Ga are best thought of as 'semi-core'.

In the DoS plots, the preceding discussion is manifested by the sharp peak in Ga-states at $(E - E_F) \sim -15$ eV coming from the 3*d* states of Ga. These semi-core states are close enough in energy to the valence states that they cannot be treated as core states and have some itinerant character, particularly at higher Ga concentrations, where an increasing overlap between the tail of the valence states and the top of the semi-core states can be seen.

When considering the states around $E_F$, we can see the relatively narrow *d*-band of Fe contrasting with the wider *sp* band coming from Ga.[1] It can be seen that the addition of Ga at low concentrations does not impact significantly the profile of the Fe *d*-bands around $E_F$. Instead, the introduced disorder can be seen to smear out the finer detail of the electronic structure. What can be seen, however, is that the tail of the Fe *d*-bands moves to lower energies as the Ga concentration increases. This

[1] A primer on the electronic structure of metals and alloys is given in Chap. 3, Sect. 3.9.

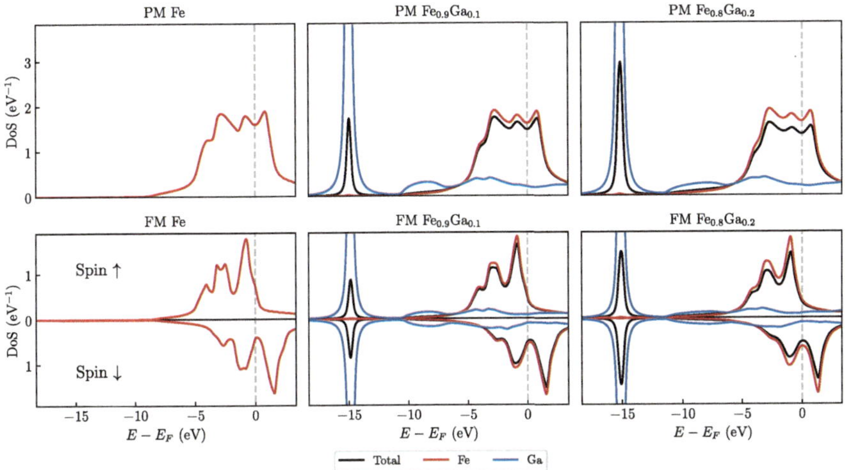


**Fig. 8.2** Species-resolved DoS for $Fe_{1-x}Ga_x$ for $x = 0, 0.1, 0.2$ in both its paramagnetic (top row) and ferromagnetic (bottom row) states. The total curve is given by the weighted average of the species-resolved curves. The sharp peak in DoS at lower energies is associated with the semi-core, localised $3d$ states of gallium

is indicative of the formation of hybridised bonding states between the $d$-like states from Fe and the $p$-like states from Ga. This hybridisation, in combination with the difference in the overall profiles of the Fe and Ga curves, we expect to result in strong ordering tendencies in this system.

## 8.4 Linear Response and Atom-Atom Interactions

Proceeding, we then perform linear response calculations for $Fe_{1-x}Ga_x$ in both its paramagnetic (DLM) and ferromagnetic states. For simplicity, we compute these interactions at a single Ga concentration, $x = 0.2$, and use the same interactions across the range of our simulations. We restrict the fit to the first two coordination shells, and the fitted parameters are tabulated in Table 8.2.

It can be seen that, in the paramagnetic (DLM) case, the interaction is dominated by the first coordination shell while, for the ferromagnetic case, the interaction has a more significant component on the second coordination shell. This is expected to have an impact on the nature of predicted chemical order.

**Table 8.2** Atom-atom interactions for $Fe_{1-x}Ga_x$ computed for $x = 0.2$ and truncated after the first two coordination shells. All in units of meV

| Paramagnetic | | | | | |
|---|---|---|---|---|---|
| $V^{(1)}_{\alpha\alpha'}$ | Fe | Ga | $V^{(2)}_{\alpha\alpha'}$ | Fe | Ga |
| Fe | 12.518 | −50.070 | Fe | 3.265 | −13.062 |
| Ga | −50.070 | 200.280 | Ga | −13.062 | 52.247 |
| Ferromagnetic | | | | | |
| $V^{(1)}_{\alpha\alpha'}$ | Fe | Ga | $V^{(2)}_{\alpha\alpha'}$ | Fe | Ga |
| Fe | 7.619 | −30.477 | Fe | 4.354 | −17.416 |
| Ga | −30.477 | 121.910 | Ga | −17.416 | 69.663 |

## 8.5 Atomistic Modelling

The three structures known to be of interest for the $Fe_{1-x}Ga_x$ in the concentration range we are considering are A2 (disordered bcc), B2, and $D0_3$. These three structures are visualised in Fig. 8.3. In the context of this system, the structures are perhaps best thought of by considering the radial distribution of neighbouring Ga atoms around a given Ga atom.[2] The A2 structure is one with complete disorder; the likelihood of finding a Ga atom as a nearest-neighbour to another Ga atom is simply the concentration of Ga in the system. The B2 structure is one where there are no Ga-Ga nearest neighbours, and instead Ga is found as a next-nearest neighbour to other Ga atoms. Finally, the $D0_3$ structure is one where Ga has lost other Ga atoms as both nearest and next-nearest neighbours; Ga-Ga neighbours are only found on the third coordination shell.

Using the atom-atom interactions of Table 8.2, we perform two sets of simulations to investigate the nature of chemical ordering in Galfenol. The first is based on the 'lattice gas' model of Chap. 2, Sect. 2.9, where we compute a set of (partial) site occupancies which minimise the free energy of the system using a gradient-based technique. This enables compositions away from stoichiometry to be modelled in relatively small systems. Using a system of 128 lattice sites, we find that orderings are always B2- or $D0_3$-like. We therefore map all site occupancies back onto the three non-equivalent sites of the $D0_3$ unit cell. Visualised in Fig. 8.4 are the determined occupancies of these three sites as a function of Ga concentration determined at a temperature of 800 K (using the ferromagnetic interactions) and 1200 K (using the paramagnetic interactions). Dashed lines show the occupancies associated with the A2, B2, and $D0_3$ structures.

When the results for $T = 800$ K are considered, using the interactions computed assuming ferromagnetic order, it can be seen that the ordering predicted for all except very low Ga concentrations is $D0_3$ like. This is indicated by Ga sitting almost exclusively on the 4a lattice site. The $D0_3$ ordering is expected based on the

[2] We could also, equivalently, consider Fe-Fe or Fe-Ga pairs.

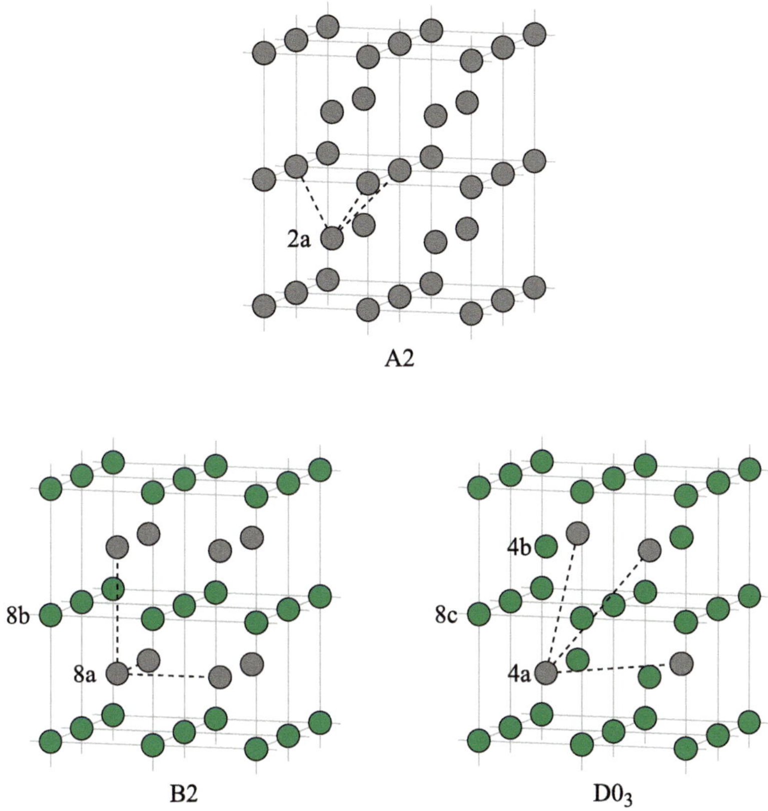


**Fig. 8.3** Visualisation of the relevant ordered structures for $Fe_{1-x}Ga_x$. A2 is disordered BCC, where all lattice sites have average occupancy $Fe_{1-x}Ga_x$. In the B2 structure, the 8b sites are pure Fe, while the 8a sites have occupancy $Fe_{1-2x}Ga_{2x}$. Finally, the $D0_3$ structure has 8c and 4c sites as pure Fe, with the 4a site occupancy given by $Fe_{1-4x}Ga_{4x}$. For clarity, sites with (partial) Ga occupancy for each structure are coloured grey. Sites are given their Wyckoff labels

longer-ranged interactions in the ferromagnetic state; a $D0_3$ ordering represents Ga-Ga pairs appearing only on the third coordination shell.

Proceeding, when the results at $T = 1200$ K are considered, using the interactions assuming a paramagnetic (DLM) state, a more mixed picture is presented. At low Ga concentrations, the predicted ordering is B2-like, then above approximately $x = 0.15$ a mixed B2/$D0_3$ state is observed. This is consistent with interactions in the paramagnetic state being dominated by the first coordination shell.

To verify the results of the simulations obtained using the lattice gas model, we perform Monte Carlo simulations and again assess the predicted atomic ordering. The simulations use a system of 1024 atoms with periodic boundary conditions applied. Each simulation is annealed from high temperature down to the target temperature in steps of 1 K, with $10^4$ Monte Carlo steps per atom performed at each temperature

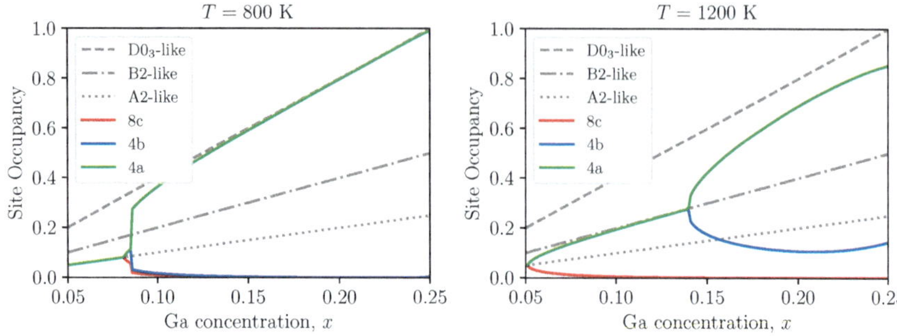


**Fig. 8.4** Average site occupancies for the three non-equivalent sites of the $D0_3$ unit cell as a function of Ga concentration at temperatures of 800 K—using the ferromagnetic interactions—and 1200 K—using the paramagnetic (DLM) interactions. The three dashed/dotted grey lines indicate the expected occupancies of perfect A2, B2, and $D0_3$ structures. Figure reproduced from [10] with permission from the American Physical Society

to ensure equilibrium is achieved. Rather than consider the Warren-Cowley SRO parameters as a function of temperature as in previous chapters, here we instead plot the radial distribution function for Ga-Ga pairs at the target temperature, and compare it to that expected for the three aforementioned ordered structures.

These are plotted in Figs. 8.5 and 8.6, for ferromagnetic and paramagnetic states respectively. The ferromagnetic simulation was equilibrated at 800 K, while the paramagnetic state was equilibrated at 1200 K. In both cases, at lower concentrations, ordering is less pronounced, as is to be expected. At higher concentrations, for the system modelled at 800 K using ferromagnetic interactions, ordering is clearly $D0_3$-like in nature. For the system modelled at 1200 K using paramagnetic interactions, the ordering is instead B2-like in nature. This verifies the results of the earlier 'lattice gas' modelling and confirms that the ordered structures of interest are B2, and $D0_3$.

These results are in agreement with existing literature. For example, a study using the coherent potential approximation and DFT calculations to directly evaluate the ordering energies of the system found that both B2 and $D0_3$ orderings were energetically favourable over the disordered A2 phase [18]. In addition, the same study reports strong ordering tendencies in the magnetically disordered state, consistent with our work. Another, later, study used *ab initio* molecular dynamics simulations to study the phase behaviour of the system and detected B2 and $D0_3$ orderings [9]. In addition, both the B2 and $D0_3$ structures are experimentally observed for this system, although we note that, at lower concentrations, it is primarily the disordered A2 phase which is detected, along with some $L1_2$ precipitates [19].

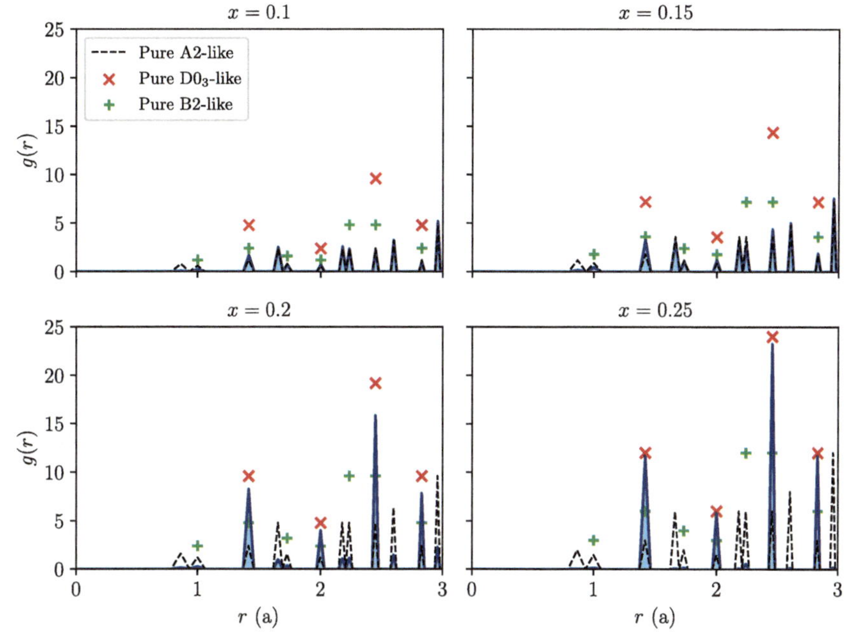


**Fig. 8.5** Radial density for Ga-Ga pairs for equilibrated simulations at 800 K using the interactions computed assuming ferromagnetic ordering, with the expected heights of peaks for A2, B2, and $D0_3$ order indicated for comparison. Results are shown for $x = 0.1, 0.15, 0.2$, and $0.25$. Particularly at higher concentrations, $D0_3$-like order can be seen emerging. Figure reproduced from [10] with permission from the American Physical Society

## 8.6 Effect of Atomic Order on Magnetic Properties

From the results of the linear response analysis and subsequent atomistic modelling, we conclude that the relevant ordered phases are B2 and $D0_3$. It remains to assess how these ordered phases impact the material's magnetic properties. For this, we will focus on the material's Curie temperature, $T_C$, and its magnetoelastic constant, $B_1$, discussed in Sect. 8.2.

### *8.6.1 Curie Temperature*

One of the most fundamental properties of of a ferromagnetic material is its Curie temperature, the temperature at which its magnetic susceptibility diverges and below which the material has non-zero magnetisation even in zero applied field. We compute the Curie temperatures of $Fe_{1-x}Ga_x$ as a function of $x$ for A2, B2 and $D0_3$ phases, using the disordered local moment (DLM) description of magnetic disorder at finite temperature [20], which is outlined in Chap. 3.

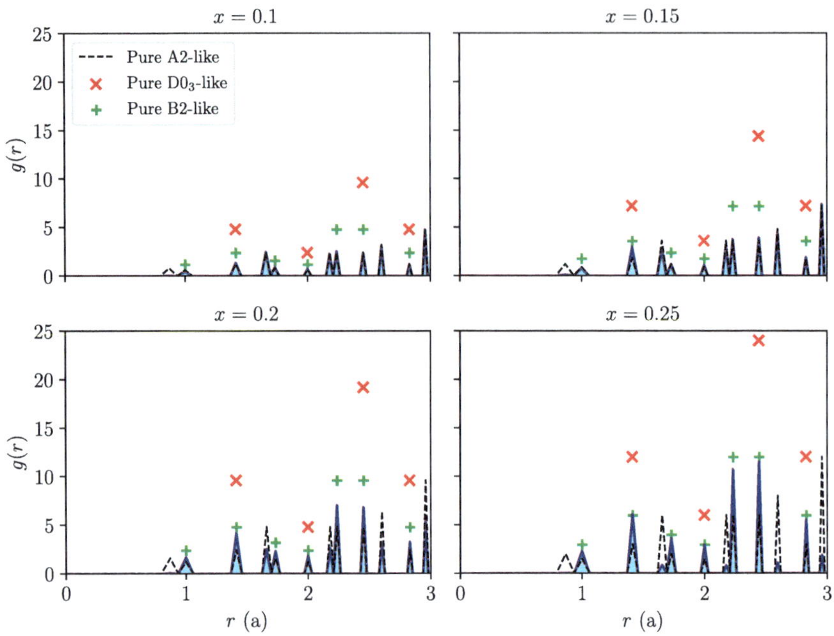


**Fig. 8.6** Radial density for Ga-Ga pairs for equilibrated simulations at 1200 K using the interactions computed assuming a paramagnetic state, with the expected heights of peaks for A2, B2, and $D0_3$ order indicated for comparison. Results are shown for $x = 0.1$, 0.15, 0.2, and 0.25. The ordering tendencies are clearly B2-like in this case

The computational approach for using DLM calculations to obtain an estimate of the Curie temperature of a system is described in detail in other works, and the reader is referred to those for full details of the underlying theory [20–24]. In essence, for a given value of $\beta = 1/(k_B T)$, the approach writes the magnetisation of a system, $\mathbf{m}_i$ in terms of self-consistently determined Weiss fields, $\mathbf{h}_i$, writing

$$\mathbf{m}_i = L(\beta \mathbf{h}_i), \tag{8.6}$$

where $L(x)$ is the Langevin function, defined as

$$L(x) = \coth(x) - \frac{1}{x}, \tag{8.7}$$

which reduces to $L(x) \approx x/3$ in the limit of small $x$. Therefore, in the limit of small $\beta \mathbf{h}_i$, 8.6 can be linearised to obtain an estimate of the Curie temperature.

For Galfenol, the Ga atoms do not support a local moment in the paramagnetic state, so it is sufficient to consider the magnetisation of the Fe atoms (appropriately weighted by concentrations). In the case of a single non-equivalent lattice site in the unit cell ($\mathbf{m}_i = \mathbf{m}$, $\mathbf{h}_i = \mathbf{h}$) the linearised equation is written [20].

$$\mathbf{m} \approx \frac{1}{3}\beta\mathbf{h}. \tag{8.8}$$

For small $\mathbf{m}$ is is reasonable to suppose that the Weiss field is linear in $\mathbf{m}$ and it is therefore possible to write

$$\mathbf{h} = J\mathbf{m}. \tag{8.9}$$

Given a few small values of $\beta\mathbf{h}$, it is possible to perform a linear fit to obtain $J$, then the Curie temperature is found as $T_C = J/(3k_B)$. This is applied to compute the Curie temperature of the atomically disordered A1 phase.

For the atomically ordered B2 and $D0_3$ phases, where there are multiple non-equivalent lattice sites in the unit cell, the same linearisation can be performed but this time a matrix equation is obtained. A more detailed discussion of this procedure can be found in [22]. For the B2 phase, for example, where there are two non-equivalent Fe sites in the unit cell, and the Weiss fields for the two non-equivalent Fe sites are written

$$\begin{pmatrix} \mathbf{h}_{8a} \\ \mathbf{h}_{8b} \end{pmatrix} \approx \begin{pmatrix} c_{\mathrm{Fe}_{8a}} J_{8a\text{-}8a} & J_{8a\text{-}8b} \\ c_{\mathrm{Fe}_{8a}} J_{8a\text{-}8b} & J_{8b\text{-}8b} \end{pmatrix} \begin{pmatrix} \mathbf{m}_{8a} \\ \mathbf{m}_{8b} \end{pmatrix}, \tag{8.10}$$

where $c_{\mathrm{Fe}_{8a}}$ is the concentration of Fe on the 8a site, and all other symbols are generalisations of the symbols of Equations 8.8 and 8.9 to the case of multiple sublattices. The Curie temperature is then obtained in the limit of small $\beta\mathbf{h}_i$ by solving the eigenvalue problem

$$\begin{pmatrix} \mathbf{h}_{8a} \\ \mathbf{h}_{8b} \end{pmatrix} = \frac{\beta}{3} \begin{pmatrix} c_{\mathrm{Fe}_{8a}} J_{8a\text{-}8a} & J_{8a\text{-}8b} \\ c_{\mathrm{Fe}_{8a}} J_{8a\text{-}8b} & J_{8b\text{-}8b} \end{pmatrix} \begin{pmatrix} \mathbf{h}_{8a} \\ \mathbf{h}_{8b} \end{pmatrix}, \tag{8.11}$$

where the smallest eigenvalue of the matrix

$$\frac{1}{3k_B} \begin{pmatrix} c_{\mathrm{Fe}_{8a}} J_{8a\text{-}8a} & J_{8a\text{-}8b} \\ c_{\mathrm{Fe}_{8a}} J_{8a\text{-}8b} & J_{8b\text{-}8b} \end{pmatrix} \tag{8.12}$$

corresponds to $T_C$. This approach can also be generalised to three magnetic sublattices, as necessary for the $D0_3$ structure.

For all three structures, the Curie temperature as a function of Ga concentration is visualised in Fig. 8.7. These results are obatined using the MARMOT code [24]. All Curie temperatures are anticipated to be overestimates of their true values, having been obtained within a single-site theory. In addition, DLM calculations can be extremely sensitive to the precise value of lattice parameter used. However, we expect that these calculations will detect qualitative trends in the magnetic behaviour of a material.

It can be seen that the emergence of atomic order reduces the Curie temperature of the material. This is understood by considering that ferromagnetism is a collective phenomenon. In $Fe_{1-x}Ga_x$, Ga is essentially non-magnetic, so any magnetic

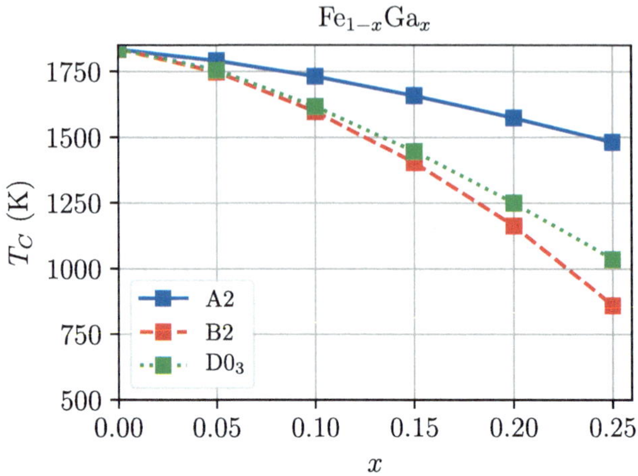


**Fig. 8.7** Plots of the predicted Curie temperature, $T_C$ of $Fe_{1-x}Ga_x$ as a function of $x$. The atomically ordered B2 and D0$_3$ phases have a reduced predicted Curie temperature compared to the atomically disordered A2 phase

behaviour has its origins in Fe. In the disordered phase, on average, an Fe atom 'sees' $8(1-x)$ neighbouring Fe atoms around it, and interacts with them to form a spontaneous magnetic moment. In the ordered B2 phase, however, an Fe atom on the Fe sublattice 'sees' $8(1-2x)$ neighbouring Fe atoms, a notable reduction on the ordered case, which results in a reduced moment on this sublattice. Although any Fe sitting on the Ga sublattice sees a large number of Fe neighbours, on average these neighbours have reduced magnetic moments, and the collective result is that a ferromagnetic state is less energetically favourable than before, and collapses at lower temperatures. A similar picture can be used to understand the behaviour of the D0$_3$ case, where the increased localisation of the Ga atoms to one of four non-equivalent lattice sites results in a fractional increase in predicted $T_C$ over the B2 structure. The predicted Curie temperatures could be reduced by applying an Onsager correction to this mean-field approach [25], similar in manner to the one applied to restore on-site sum rules for the alloys in Chap. 2, but we omit such calculations here.

### *8.6.2 Magnetoelastic Constants*

As mentioned in the introduction to this chapter, a primary magnetic quantity of interest for Galfenol is its magnetostriction and the magnetoelastic constants quantifying this. Within the framework outlined in Sect. 8.2, George A. Marchant has performed calculations to obtain the magnetoelastic constant $B_1$ for the three considered structures as a function of Ga concentration [10]. Plots of $B_1$ vs Ga concentration, $x$, are visualised in Fig. 8.8.

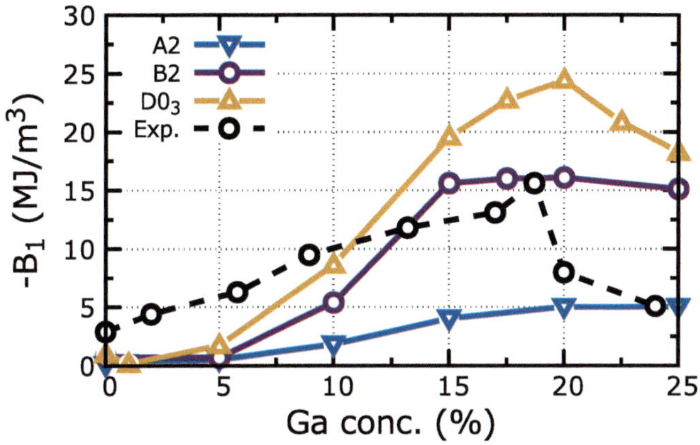


**Fig. 8.8** Plots of the predicted magnetoelastic constant, $B_1$, of $Fe_{1-x}Ga_x$ as a function of $x$ for the A2, B2, and $D0_3$ structures. Figure courtesy of Dr G. A. Marchant

It can be seen that modelling the system in the A2 phase fails to reproduce (either qualitatively or quantitatively) the experimentally observed dependence of the magnetoelastic constant on Ga concentration. However, both the B2 and $D0_3$ ordered phases produce values of $B_1$ which are quantitatively closer to experiment, while the $D0_3$ phase even successfully reproduces the experimentally observed peak in $B_1$ at $x \approx 0.19$. A full discussion of these results, as well as discussion of their origins in terms of the material's underlying electronic structure, can be found in References [6, 10].

## 8.7 Summary

In summary, we have modelled the phase behaviour of the system $Fe_{1-x}Ga_x$ in the range $0 < x \leq 0.25$. The atomically ordered phases of interest have been confirmed to be the B2 and $D0_3$ structures. We find that in the paramagnetic state, B2 ordering is preferred, while in the ferromagnetic state, longer-ranged interactions favour $D0_3$ ordering. These results suggest that a sample annealed above its Curie temperature may preferentially order into the B2 phase, while a sample annealed below its Curie temperature (or in an applied magnetic field) may preferentially form a $D0_3$ structure.

We have then studied the effect of these chemical orderings on the material's magnetic properties. It is found that the Curie temperature of the B2 and $D0_3$ atomically ordered structures are predicted to be reduced compared to the atomically disordered A2 phase. In addition, the predicted magnetoelastic constants for the three phases (A2, B2, $D0_3$) have been examined, and it is found that the B2 and $D0_3$ ordered structures yield magnetoelastic constants which compare much more favourably with experiment than the disordered A2 phase.

These results demonstrate that it is essential to obtain a complete description of atomic arrangements in an alloy before considering other materials properties. They therefore outline how the approach for modelling atomic arrangements in multicomponent alloys outlined in this work can be used for materials design when combined with modelling techniques capable of modelling the impact of atomic ordering on physical properties of a system.

## References

1. A.E. Clark, K.B. Hathaway, M. Wun-Fogle, J.B. Restorff, T.A. Lograsso, V.M. Keppens, G. Petculescu, R.A. Taylor, J. Appl. Phys. **93**(10), 8621 (2003). https://doi.org/10.1063/1.1540130
2. J. Atulasimha, A.B. Flatau, Smart Mater. Struct. **20**(4), 043001 (2011). https://doi.org/10.1088/0964-1726/20/4/043001
3. R. Grössinger, R.S. Turtelli, N. Mehmood, I.O.P. Conf, Ser. Mater. Sci. Eng. **60**, 012002 (2014). https://doi.org/10.1088/1757-899x/60/1/012002
4. E.d.T.d. Lacheisserie, *Magnetostriction: Theory and Applications of Magnetoelasticity* (CRC Press, 1993)
5. N. Srisukhumbowornchai, S. Guruswamy, J. Appl. Phys. **90**(11), 5680 (2001). https://doi.org/10.1063/1.1412840
6. G.A. Marchant, C.E. Patrick, J.B. Staunton, Phys. Rev. B **99**(5), 054415 (2019). https://doi.org/10.1103/PhysRevB.99.054415
7. Y.N. Zhang, J.X. Cao, R.Q. Wu, Appl. Phys. Lett. **96**(6), 062508 (2010). https://doi.org/10.1063/1.3318420
8. R. Wu, J. Appl. Phys. **91**(10), 7358 (2002). https://doi.org/10.1063/1.1450791
9. H. Wang, Y.N. Zhang, R.Q. Wu, L.Z. Sun, D.S. Xu, Z.D. Zhang, Sci. Rep. **3**(1), 3521 (2013). https://doi.org/10.1038/srep03521
10. G.A. Marchant, C.D. Woodgate, C.E. Patrick, J.B. Staunton, Phys. Rev. B **103**(9), 094414 (2021). https://doi.org/10.1103/PhysRevB.103.094414
11. C. Kittel, Rev. Mod. Phys. **21**(4), 541 (1949). https://doi.org/10.1103/RevModPhys.21.541
12. A. Clark, J. Restorff, M. Wun-Fogle, T. Lograsso, D. Schlagel, IEEE Trans. Magn. **36**(5), 3238 (2000). https://doi.org/10.1109/20.908752
13. C. Dasarathy, W. Hume-Rothery, Proceed. R. Soc. London. Ser. A. Math. Phys. Sci. **286**(1405), 141 (1965). http://www.jstor.org/stable/2415280
14. Z.S. Basinski, W. Hume-Rothery, A.L. Sutton, Proceed. R. Soc. London. Ser. A. Math. Phys. Sci. **229**(1179), 459 (1955). http://www.jstor.org/stable/99693
15. M. Hoffmann, A. Ernst, W. Hergert, V.N. Antonov, W.A. Adeagbo, R.M. Geilhufe, H. Ben Hamed, Physica Status Solidi (b) **257**(7), 1900671 (2020). https://doi.org/10.1002/pssb.201900671
16. G.M. Stocks, W.M. Temmerman, B.L. Gyorffy, Phys. Rev. Lett. **41**(5), 339 (1978). https://doi.org/10.1103/PhysRevLett.41.339
17. J.P. Perdew, Y. Wang, Phys. Rev. B **45**(23), 13244 (1992). https://doi.org/10.1103/PhysRevB.45.13244
18. T. Khmelevska, S. Khmelevskyi, P. Mohn, J. Appl. Phys. **103**(7), 073911 (2008). https://doi.org/10.1063/1.2903071
19. A. Mohamed, V. Palacheva, V. Cheverikin, E. Zanaeva, W. Cheng, V. Kulitckii, S. Divinski, G. Wilde, I. Golovin, J. Alloys Comp. **846**, 156486 (2020). https://doi.org/10.1016/j.jallcom.2020.156486
20. B.L. Gyorffy, A.J. Pindor, J. Staunton, G.M. Stocks, H. Winter, J. Phys. F: Metal Phys. **15**(6), 1337 (1985). https://doi.org/10.1088/0305-4608/15/6/018

21. C.E. Patrick, S. Kumar, G. Balakrishnan, R.S. Edwards, M.R. Lees, E. Mendive-Tapia, L. Petit, J.B. Staunton, Phys. Rev. Mater. **1**(2), 024411 (2017). https://doi.org/10.1103/PhysRevMaterials.1.024411
22. K. Nawa, I. Kurniawan, K. Masuda, Y. Miura, C.E. Patrick, J.B. Staunton, Phys. Rev. B **102**(5), 054424 (2020). https://doi.org/10.1103/PhysRevB.102.054424
23. C.E. Patrick, S. Kumar, G. Balakrishnan, R.S. Edwards, M.R. Lees, L. Petit, J.B. Staunton, Phys. Rev. Lett. **120**(9), 097202 (2018). https://doi.org/10.1103/PhysRevLett.120.097202
24. C.E. Patrick, J.B. Staunton, Electron. Struct. **4**(1), 017001 (2022). https://doi.org/10.1088/2516-1075/ac5755
25. J.B. Staunton, B.L. Gyorffy, Phys. Rev. Lett. **69**(2), 371 (1992). https://doi.org/10.1103/PhysRevLett.69.371

# Chapter 9
# Summary, Conclusions, and Outlook

**Abstract** We give an overview of the modelling approach presented in this book, and summarise the results obtained by applying it to a number of prototypical multicomponent alloys. We then draw conclusions, before giving an outlook on future possibilities and avenues of study.

## 9.1 Summary and Conclusions

In summary, an approach for modelling atomic arrangements in multicomponent alloys has been outlined, implemented, and applied to a number of example systems. The method is based on a Taylor expansion of the Gibbs free energy of the disordered solid solution, where the free energy is obtained via first-principles density functional theory calculations. Through application of a Landau-type linear response theory, it is possible to infer disorder-order transitions directly and characterise the structure of (partially) ordered phases. In addition, the linear response theory can be used to obtain pairwise atom-atom interaction energies suitable for lattice-based atomistic modelling. This allows the nature of atomic arrangements at temperatures below any initial ordering temperature to be assessed, and also provides clearer understanding of atomic ordering tendencies. Further, the theory enables elucidation of ordering tendencies in terms of underlying physical mechanisms, which a purely data-driven approach would fail to do. Finally, it is noted that the method is exceptionally computationally efficient; for a given multicomponent system it is possible to perform a comprehensive analysis using $\sim$100 CPU hours on a modern computing cluster.

The first group of systems to which the method has been applied are the refractory high-entropy alloys [1], a family of multicomponent alloys forming on the bcc lattice containing the elements V, Nb, Mo, Ta, and W. The results obtained gave physical insight into the origins of atomic ordering, with the crucial mechanisms being atomic size/bandwidth differences and valence differences between species. Both the nature and predicted temperatures associated with atomic short- and long-range ordering were consistent with existing literature, validating the robustness of the approach.

Proceeding, the methodology was then used to study the phase behaviour of the $Ti_x$NbMoTaW and $Ti_x$VNbMoTaW high-entropy alloys [2], which are



C. D. Woodgate, *Modelling Atomic Arrangements in Multicomponent Alloys*,
Springer Series in Materials Science 346,
https://doi.org/10.1007/978-3-031-62021-8_9

technologically relevant as the addition of Ti improves these systems' room-temperature ductility. In alignment with the experimental data, we find that these systems exhibit competition between phase segregation and phase ordering, leading to emergent multiphase behaviour at intermediate temperatures. This is understood to originate from the strong interactions between Ti and the other elements on account of atomic size, bandwidth, and valence differences.

In a progression, we then considered alloys containing magnetic elements: the Cantor-Wu medium- and high-entropy alloys, a family of multicomponent alloys forming on the fcc lattice [3]. These systems were modelled in the paramagnetic state via application of the disordered local moment picture. The results obtained again give intuition as to the underlying physical origins of atomic ordering; correlations between Cr, Co, and Ni are dominant, with Fe and Mn serving a diluting effect and stabilising the solid solution down to low temperatures. Once again, the obtained results compare well with existing literature, further validating the approach.

To elucidate the importance of correct treatment of the magnetic state when modelling atomic arrangements in alloys, the Cantor-Wu systems have been analysed again, but this time using electronic structure calculations describing both non-magnetic and ferrimagnetic states, to compare with the results provided for modelling of the paramagnetic state [4]. The non-magnetic state is unphysical for these systems, which contain Fe and Mn, elements which will remain paramagnetic to high temperatures. By contrast, the ferrimagnetic state is physically meaningful, but only emerges at low temperatures, well below the temperatures at which alloys are processed. The results for atomic short- and long-range order when modelling the Cantor-Wu systems as either ferrimagnetic or non-magnetic differ both qualitatively and quantitatively from the results for the paramagnetic state, emphasising that correct treatment of a system's magnetic state is essential to obtain physical results when modelling systems containing magnetic elements.

Then, to illustrate the connections between atomic arrangements and magnetic properties, we study the $Fe_{1-x}Ga_x$ system, commonly referred to as Galfenol [5]. The predicted ordered structures on the bcc lattice for this system are B2 and $D0_3$. Results have been presented which illustrate that this chemical ordering impacts the Curie temperature of the system, and also its magnetoelastic constant. The $D0_3$ phase in particular seems to exhibit predicted magnetoelastic behaviour consistent with the experimental data. These results serve to emphasise the important connections between atomic arrangements and observable physical properties in alloy systems.

## 9.2 Outlook

Moving forward, there are a number of avenues of further work that could be of interest. The first, most elementary step would be to extend the approach and associated software to handle hexagonal close-packed systems with a two-atom basis, but in which both lattice sites are crystalographically equivalent. There are a number of known HEAs which form on this lattice type and the ability to model them would be of value to the materials science community. In addition to this line of work, the formalism should also be extensible to studying the phase behaviour of two-dimensional alloys (films) and also to systems in which one sublattice could be considered a solid solution, such as the high-entropy oxides and high-entropy carbides.

In a direct connection with alternative modelling approaches, we note that the computational efficiency of our lattice-based approach could be used to provide physically meaningful configurations for use in more computationally expensive approaches. For example, training a machine-learned interatomic potential for an alloy requires configurations that are both disordered and also those that contain a degree of atomic short-range order. Ongoing work suggests that the lattice-based configurations generated using the approach outlined in this work can be used as training data for interatomic potentials and represent an improvement over configurations generated using other methods such as special quasirandom structures [6].

Finally, and most ambitiously, there is clearly scope to use this method for materials design and discovery. The low computational cost means that it would not be unreasonable to screen hundreds or thousands of potential alloy configurations to search for new intermetallic phases (characterised by a high atomic ordering temperature) or high-entropy compositions (characterised by a low atomic ordering temperature). The method has the potential to screen a wide range of compositions and narrow down candidate materials for further computational/experimental investigation.

## 9.3 A Message to Early-Career Scientists

Before I finish, I will add a short message to any early-career scientists reading this book. Science and, in particular, computational science is *hard*. Very often it feels like there is too much to learn, and not enough time to learn it! If it is of any help, I would point you to a number of resources prepared by myself and other early-career scientists with whom I have worked. Much of this work was spearheaded by Dr. Idil Ismail (formerly Univeristy of Warwick, now ETH Zurich), to whom we are all indebted for her hard work preparing these resources. We can offer:

- A podcast series aimed at those interested in computational science, titled 'Multiscale Musings': https://warwick.ac.uk/fac/sci/hetsys/outreach/multiscalemusings/
- A seminar series aimed at giving early-career researchers the computational tools they need to succeed, titled 'The Computational Toolkit': https://computationaltoolkit.github.io/
- A 'Tutorial' article aimed as a guide for graduate students entering the field of computational science [7]: Ismail *et al.*, 'Eat, sleep, code, repeat: tips for early-career researchers in computational science', Eur. Phys. J. Plus. **138**, 1094 (2023).

It is my hope that some or all of these may be of use to you.

## References

1. C.D. Woodgate, J.B. Staunton, Phys. Rev. Mater. **7**(1), 013801 (2023). https://doi.org/10.1103/PhysRevMaterials.7.013801
2. C.D. Woodgate, J.B. Staunton, J. Appl. Phys. **135**(13), 135106 (2024). https://doi.org/10.1063/5.0200862
3. C.D. Woodgate, J.B. Staunton, Phys. Rev. B **105**(11), 115124 (2022). https://doi.org/10.1103/PhysRevB.105.115124
4. C.D. Woodgate, D. Hedlund, L.H. Lewis, J.B. Staunton, Phys. Rev. Mater. **7**(5), 053801 (2023). https://doi.org/10.1103/PhysRevMaterials.7.053801
5. G.A. Marchant, C.D. Woodgate, C.E. Patrick, J.B. Staunton, Phys. Rev. B **103**(9), 094414 (2021). https://doi.org/10.1103/PhysRevB.103.094414
6. L. Shenoy, C.D. Woodgate, J.B. Staunton, A.P. Bartók, C.S. Becquart, C. Domain, J.R. Kermode, Phys. Rev. Mater. **8**(3), 033804 (2024). https://doi.org/10.1103/PhysRevMaterials.8.033804
7. I. Ismail, S. Chaudhuri, D. Morgan, C.D. Woodgate, Z. Fakhoury, J.M. Targett, C. Pilgrim, C. Maino, European Phys. J. Plus **138**(12), 1094 (2023). https://doi.org/10.1140/epjp/s13360-023-04732-5

# Appendix A
# Special Points of Brillouin Zones

When visualising Bloch spectral functions, bandstructures, and eigenvalues of chemical stability matrices, we are really viewing quantities which are functions of a three-dimensional wave-vector **k**. (Indeed, we can often plot these quantities 'sliced' through some 2D region of reciprocal space.) By translational symmetry, we only need consider those wavevectors which cannot be related to other wavevectors by symmetry operations of the lattice, which leads us to considering the region of reciprocal space referred to as the Irreducible Brillouin Zone (IBZ), for which an excellent introduction can be found in textbooks such as [1] or [2].

The conventional wave of visualising quantities around the IBZ is to take a path in reciprocal space around high-symmetry points of the reciprocal lattice, as these are often most relevant to the physics. For reference, the conventional labels for the high-symmetry points of the bcc and fcc lattices are tabulated in Tables A.1 and A.2 respectively. We also provide the special points of the IBZ of the simple tetragonal lattice in Table A.3, to assist in comprehension of some of the BSF plots in Chap. 3. In addition, we note that an impressively comprehensive list of special points of Brillouin zones for a range of common lattices is given in [3].

**Table A.1** Special points of the IBZ of the bcc lattice

| Label | Wavevector $(2\pi/a)$ |
|---|---|
| $\Gamma$ | (0, 0, 0) |
| $H$ | (0, 0, 1) |
| $P$ | (1/2, 1/2, 1/2) |
| $N$ | (0, 1/2, 1/2) |



C. D. Woodgate, *Modelling Atomic Arrangements in Multicomponent Alloys*, Springer Series in Materials Science 346,
https://doi.org/10.1007/978-3-031-62021-8

**Table A.2** Special points of the IBZ of the fcc lattice

| Label | Wavevector ($2\pi/a$) |
|---|---|
| $\Gamma$ | (0, 0, 0) |
| $X$ | (0, 1, 0) |
| $L$ | (1/2, 1/2, 1/2) |
| $W$ | (1/2, 1, 0) |
| $U$ | (1/4, 1, 1/4) |
| $K$ | (3/2, 3/2, 0) |

**Table A.3** Special points of the IBZ of the simple tetragonal lattice, where $c \neq a$

| Label | Wavevector |
|---|---|
| $\Gamma$ | (0, 0, 0) |
| $X$ | $(\pi/a, 0, 0)$ |
| $M$ | $(\pi/a, \pi/a, 0)$ |
| $Z$ | $(0, 0, \pi/c)$ |
| $R$ | $(\pi/a, 0, \pi/c)$ |
| $A$ | $(\pi/a, \pi/a, \pi/c)$ |

## References

1. C. Kittel, *Introduction to Solid State Physics*, 8th edn. (Wiley, Hoboken, NJ, 2005)
2. S.H. Simon, *The Oxford Solid State Basics*, 1st edn. (Oxford University Press, Oxford, 2013)
3. W. Setyawan, S. Curtarolo, Comput. Mater. Sci. **49**(2), 299 (2010). https://doi.org/10.1016/j.commatsci.2010.05.010

# Appendix B
# Magnetocrystalline Anisotropy *Ab Initio*

As outlined in Chap. 8, the magnetocrystalline anisotropy of a system describes the variation in energy of the system due to a rotation of the magnetisation direction. In a tetragonal, uniaxial ferromagnet, the MAE is well-described by [1]

$$K = K_1 \sin^2\theta + K_2 \sin^4\theta + K_3 \sin^4\theta \cos 4\phi, \tag{B.1}$$

where $K_1$, $K_2$ and $K_3$ are the conventional anisotropy constants. The total energy of the system is then written as

$$\mathcal{E}(\hat{\mathbf{n}}) = \mathcal{E}_{\text{iso}} + K_1 \sin^2\theta + K_2 \sin^4\theta + K_3 \sin^4\theta \cos 4\phi, \tag{B.2}$$

where $\mathcal{E}_{\text{iso}}$ represents the isotropic portion of the energy. The MAE coefficients $K_1$, $K_2$, and $K_3$ are accessed from derivatives of this energy with respect to magnetisation directions. Given an energy of the form in (B.2), we have that

$$\begin{aligned} \frac{\partial \mathcal{E}}{\partial \theta} &= K_1 \sin 2\theta + 2K_2 \sin 2\theta \sin^2\theta \\ &\quad + 2K_3 \sin 2\theta \sin^2\theta \cos 4\phi, && \text{(B.3)} \\ \frac{\partial \mathcal{E}}{\partial \phi} &= -4K_3 \sin^4\theta \sin 4\phi, && \text{(B.4)} \end{aligned}$$

and evaluation of these quantities at particular values of $\theta$ and $\phi$ leads directly to the constants.

To evaluate these derivatives and, therefore, the MAE coefficients, we use an expression from the KKR multiple scattering formalism for the magnetic torque [2–5],

$$\mathbf{T}^{(\hat{\mathbf{n}})} = -\frac{\partial \mathcal{E}^{(\hat{\mathbf{n}})}}{\partial \hat{\mathbf{n}}}, \tag{B.5}$$



C. D. Woodgate, *Modelling Atomic Arrangements in Multicomponent Alloys*, Springer Series in Materials Science 346,
https://doi.org/10.1007/978-3-031-62021-8

in which the effects of finite temperature local moment fluctuations [6] can also be included. Using the stationary properties of the relativistic DFT energy $\mathcal{E}^{(\hat{\mathbf{n}})}$ with respect to the charge and magnetisation densities, the derivation begins with the single-electron energy-sum part of this energy. In terms of the integrated electronic density of states, $N^{(\hat{\mathbf{n}})}(E)$, for a system magnetised along a direction $\hat{\mathbf{n}}$, this energy is written as

$$\mathcal{E}^{(\hat{\mathbf{n}})} = -\int^{E_F^{(\hat{\mathbf{n}})}} N^{(\hat{\mathbf{n}})}(E)\,\mathrm{d}E. \tag{B.6}$$

In multiple scattering theory the integrated density of states is expressed particularly succinctly using the Lloyd formula [7, 8]

$$N^{(\hat{\mathbf{n}})}(E) = N_0(E) - \frac{1}{\pi}\,\mathrm{Im}\ln\det\left(\underline{\underline{t}}(\hat{\mathbf{n}};E)^{-1} - \underline{\underline{G}}_0(E)\right), \tag{B.7}$$

where $\underline{\underline{t}}(\hat{\mathbf{n}};E)$ describes an array of single-site-scattering t-matrices (combined into a super matrix in both site and angular momentum space) and $\underline{\underline{G}}_0(E)$ specifies the structure constants which contain all the information on the spatial location of the scatterers [8].

At a site $i$, the t-matrix describing the scattering of an electron from an effective scalar potential and local magnetic field, aligned with the spin-polarisation of the electron density in that region and located in the unit cell surrounding the site, is obtained from the solution of the Dirac equation. This solution is found in a coordinate frame in which the z-axis of the local coordinate frame is aligned with $\hat{\mathbf{n}}$ [9, 10]. A simple transformation produces a t-matrix for the general coordinate frame,

$$\underline{t}_i(\hat{\mathbf{n}};E) = \underline{R}(\hat{\mathbf{n}})\,\underline{t}_i(\hat{\mathbf{z}};E)\underline{R}(\hat{\mathbf{n}})^{+} \tag{B.8}$$

where $\underline{R}(\hat{\mathbf{n}}) = \exp i\alpha_{\hat{\mathbf{m}}}(\hat{\mathbf{m}}\cdot\underline{\mathbf{J}})$, $\alpha_{\hat{\mathbf{m}}}$ is the angle of rotation about an axis $\hat{\mathbf{m}} = \frac{\hat{\mathbf{z}}\wedge\hat{\mathbf{n}}}{|\hat{\mathbf{z}}\wedge\hat{\mathbf{n}}|}$, and $\underline{\mathbf{J}}$ is the total angular momentum. The torque quantity $T^{(\hat{\mathbf{n}})}_{\alpha_{\hat{u}}} = -\frac{\partial\mathcal{E}^{(\hat{\mathbf{n}})}}{\partial\alpha_{\hat{\mathbf{u}}}}$, describing the variation of the total energy with respect to a rotation of the magnetisation about a general axis $\hat{\mathbf{u}}$, is

$$T^{(\hat{\mathbf{n}})}_{\alpha_{\hat{\mathbf{u}}}} = -\frac{1}{\pi}\int^{E_F^{(\hat{\mathbf{n}})}} \mathrm{Im}\,\frac{\partial}{\partial\alpha_{\hat{\mathbf{u}}}}\left[\ln\det\left(\underline{\underline{t}}(\hat{\mathbf{n}};E)^{-1} - \underline{\underline{G}}_0(E)\right)\right]\mathrm{d}E \tag{B.9}$$

which can be written as [3]

$$T^{(\hat{\mathbf{n}})}_{\alpha_{\hat{\mathbf{u}}}} = -\frac{1}{\pi}\int^{E_F^{(\hat{\mathbf{n}})}} \mathrm{Im}\sum_i \mathrm{Tr}\left(\underline{\tau}^{(\hat{\mathbf{n}})}_{ii}(E)\frac{\partial}{\partial\alpha_{\hat{\mathbf{u}}}}\left(\underline{R}(\hat{\mathbf{n}})\underline{t}(\hat{\mathbf{n}};E)^{-1}\underline{R}(\hat{\mathbf{n}})^{+}\right)\right)\mathrm{d}E \tag{B.10}$$

where the KKR scattering path operator [11, 8] is

$$\underline{\underline{\tau}}^{(\hat{\mathbf{n}})} = \left( \left( \underline{\underline{t}}^{(\hat{\mathbf{n}})} \right)^{-1} - \underline{\underline{G}}_0 \right)^{-1} . \tag{B.11}$$

Since $\frac{\partial \underline{R}(\hat{\mathbf{n}})}{\partial \alpha_{\hat{\mathbf{u}}}} = i(\underline{\mathbf{J}} \cdot \hat{\mathbf{u}})\underline{R}(\hat{\mathbf{n}})$ and $\frac{\partial \underline{R}(\hat{n})^{+}}{\partial \alpha_{\hat{\mathbf{u}}}} = -i(\underline{\mathbf{J}} \cdot \hat{\mathbf{u}})\underline{R}(\hat{\mathbf{n}})$, we obtain

$$T_{\alpha_{\hat{\mathbf{u}}}}^{(\hat{\mathbf{n}})} = \frac{1}{\pi} \int^{E_F^{(\hat{\mathbf{n}})}} \operatorname{Im} i \sum_i \operatorname{Tr} \left( \underline{\tau}_{ii}^{(\hat{\mathbf{n}})}(E) \left[ (\underline{\mathbf{J}} \cdot \hat{\mathbf{u}})\, \underline{t}(\hat{\mathbf{n}}; E)^{-1} - \underline{t}(\hat{\mathbf{n}}; E)^{-1} (\underline{\mathbf{J}} \cdot \hat{\mathbf{u}}) \right] \right) \mathrm{d}E. \tag{B.12}$$

This closed-form expression enables us to evaluate the torque for a given magnetisation direction $\hat{\mathbf{n}}$. By sampling a range of values of $\theta$ and $\phi$, it is therefore possible to extract values of the coefficients $K_1$, $K_2$, and $K_3$ separately along with the conventional uniaxial anisotropy, $K_U$. We note that this technique for evaluating the MAE has previously exhibited exceptional fidelity when applied to other alloys crystallising in the $L1_0$ structure, giving values in excellent agreement with experimental literature for the FePd [3] and FePt [2] systems, along with explaining the dependence of the MAE on the long-range atomic order parameter [12]. In addition, the method has also been used with success to study a number of rare-earth-based ferromagnetic systems [13, 14] and other magnetic materials [15].

## References


1. L.D. Landau, E.M. Lifshitz, L.P. Pitaevskii, *Electrodynamics of Continuous Media*, 2nd edn. No. 8 in Course of Theoretical Physics (Elsevier Butterworth-Heinemann, Amsterdam Heidelberg, 2009)
2. J.B. Staunton, S. Ostanin, S.S.A. Razee, B.L. Gyorffy, L. Szunyogh, B. Ginatempo, E. Bruno, Phys. Rev. Lett. **93**(25), 257204 (2004). https://doi.org/10.1103/PhysRevLett.93.257204
3. J.B. Staunton, L. Szunyogh, A. Buruzs, B.L. Gyorffy, S. Ostanin, L. Udvardi, Phys. Rev. B **74**(14), 144411 (2006). https://doi.org/10.1103/PhysRevB.74.144411
4. C.E. Patrick, J.B. Staunton, Electron. Struct. **4**(1), 017001 (2022). https://doi.org/10.1088/2516-1075/ac5755
5. S. Ouazi, S. Vlaic, S. Rusponi, G. Moulas, P. Buluschek, K. Halleux, S. Bornemann, S. Mankovsky, J. Minár, J. Staunton, H. Ebert, H. Brune, Nat. Commun. **3**(1), 1313 (2012). https://doi.org/10.1038/ncomms2316
6. B.L. Gyorffy, A.J. Pindor, J. Staunton, G.M. Stocks, H. Winter, J. Phys. F: Metal Phys. **15**(6), 1337 (1985). https://doi.org/10.1088/0305-4608/15/6/018
7. P. Lloyd, Proc. Phys. Soc. **90**(1), 207 (1967). https://doi.org/10.1088/0370-1328/90/1/323, https://iopscience.iop.org/article/10.1088/0370-1328/90/1/323
8. J.S. Faulkner, G.M. Stocks, Y. Wang, *Multiple Scattering Theory: Electronic Structure of Solids*, 1st edn. (IOP Publishing, Bristol, UK, 2018). https://doi.org/10.1088/2053-2563/aae7d8
9. P. Strange, J. Staunton, B.L. Gyorffy, J. Phys. C: Solid State Phys. **17**(19), 3355 (1984). https://doi.org/10.1088/0022-3719/17/19/011
10. H. Ebert, D. Ködderitzsch, J. Minár, Rep. Progress Phys. **74**(9), 096501 (2011). https://doi.org/10.1088/0034-4885/74/9/096501

11. B.L. Gyorffy, M.J. Stott, *Band Structure Spectroscopy of Metals and Alloys* (Academic Press, London; New York, 1973)
12. J.B. Staunton, S. Ostanin, S.S.A. Razee, B. Gyorffy, L. Szunyogh, B. Ginatempo, E. Bruno, J. Phys.: Condens. Matter **16**(48), S5623 (2004). https://doi.org/10.1088/0953-8984/16/48/019
13. C.E. Patrick, J.B. Staunton, Phys. Rev. Mater. **3**(10), 101401 (2019). https://doi.org/10.1103/PhysRevMaterials.3.101401
14. J. Bouaziz, C.E. Patrick, J.B. Staunton, Phys. Rev. B **107**(2), L020401 (2023). https://doi.org/10.1103/PhysRevB.107.L020401
15. C.E. Patrick, Phys. Rev. B **106**(1), 014425 (2022). https://doi.org/10.1103/PhysRevB.106.014425

# Appendix C
# Atom-Atom Interchange Parameters

For reference, the atom-atom interchange parameters (effective pair interactions) fitted to the $S^{(2)}_{\alpha\alpha'}(\mathbf{k})$ data for the multicomponent systems considered in this work are explicitly tabulated here.

Atom-atom interchange parameters for the refractory HEAs of Chap. 4, obtained from $S^{(2)}$s evaluated at LDA-optimised lattice spacings, are given in Tables C.1, C.2, and C.3.

Atom-atom interchange parameters for the $Ti_x$NbMoTaW and $Ti_x$VNbMoTaW refractory HEAs of Chap. 5, obtained from $S^{(2)}$s evaluated at lattice spacings taken from experimental data, are given in Tables C.4, C.5, C.6, C.7, C.8 and C.9.

For the Cantor-Wu systems of Chaps. 6 and 7, Tables C.10, C.11, C.12, C.13, C.14, C.15, C.16 and C.17 provide atom-atom interchange parameters for the considered multicomponent systems in their non-magnetic, paramagnetic, and ferrimagnetic states.

A link to the repository containing these parameters and other data associated with the work of this thesis is given in Appendix D.



C. D. Woodgate, *Modelling Atomic Arrangements in Multicomponent Alloys*, Springer Series in Materials Science 346,
https://doi.org/10.1007/978-3-031-62021-8

**Table C.1** Atom-atom pairwise interactions for the three-component NbMoTa refractory medium-entropy alloy considered in Chap. 4, fitted from $S^{(2)}$s evaluated at $T = 1200$ K for the specified equiatomic composition. The lattice parameter used in the $S^{(2)}$ calculation is the LDA-optimised parameter of $a = 3.173$ Å. All interaction values are in units of meV. Interactions are dominated by the first two shells

| $V^{(1)}_{\alpha\alpha'}$ | Nb | Mo | Ta |
|---|---|---|---|
| Nb | 2.6 | −8.6 | 6.0 |
| Mo | −8.6 | 19.4 | −10.9 |
| Ta | 6.0 | −10.9 | 4.9 |
| $V^{(2)}_{\alpha\alpha'}$ | Nb | Mo | Ta |
| Nb | 2.0 | −2.5 | 0.5 |
| Mo | −2.5 | 3.3 | −0.8 |
| Ta | 0.5 | −0.8 | 0.3 |
| $V^{(3)}_{\alpha\alpha'}$ | Nb | Mo | Ta |
| Nb | −0.1 | −0.1 | 0.2 |
| Mo | −0.1 | 1.7 | −1.6 |
| Ta | 0.2 | −1.6 | 1.5 |
| $V^{(4)}_{\alpha\alpha'}$ | Nb | Mo | Ta |
| Nb | −0.1 | 0.1 | −0.0 |
| Mo | 0.1 | −0.2 | 0.1 |
| Ta | −0.0 | 0.1 | −0.1 |
| $V^{(5)}_{\alpha\alpha'}$ | Nb | Mo | Ta |
| Nb | 0.0 | −0.0 | 0.0 |
| Mo | −0.0 | 1.3 | −1.3 |
| Ta | 0.0 | −1.3 | 1.3 |
| $V^{(6)}_{\alpha\alpha'}$ | Nb | Mo | Ta |
| Nb | −0.3 | 0.3 | 0.1 |
| Mo | 0.3 | −0.5 | 0.2 |
| Ta | 0.1 | 0.2 | −0.2 |

**Table C.2** Atom-atom pairwise interactions for the four-component NbMoTaW refractory high-entropy alloy considered in Chap. 4, fitted from $S^{(2)}$s evaluated at $T = 1200$ K for the specified equiatomic composition. The lattice parameter used in the $S^{(2)}$ calculation is the LDA-optimised parameter of $a = 3.151$ Å. All interaction values are in units of meV. Interactions are dominated by the first two shells

| $V^{(1)}_{\alpha\alpha'}$ | Nb | Mo | Ta | W |
|---|---|---|---|---|
| Nb | 6.9 | −12.8 | 12.7 | −6.8 |
| Mo | −12.8 | 12.6 | −13.6 | 13.8 |
| Ta | 12.7 | −13.6 | 14.3 | −13.4 |
| W | −6.8 | 13.8 | −13.4 | 6.4 |
| $V^{(2)}_{\alpha\alpha'}$ | Nb | Mo | Ta | W |
| Nb | 7.1 | −2.3 | 3.2 | −8.0 |
| Mo | −2.3 | 2.6 | −1.8 | 1.6 |
| Ta | 3.2 | −1.8 | 1.5 | −2.9 |
| W | −8.0 | 1.6 | −2.9 | 9.3 |
| $V^{(3)}_{\alpha\alpha'}$ | Nb | Mo | Ta | W |
| Nb | 0.3 | −1.2 | 1.1 | −0.2 |
| Mo | −1.2 | 2.6 | −2.5 | 1.2 |
| Ta | 1.1 | −2.5 | 2.5 | −1.1 |
| W | −0.2 | 1.2 | −1.1 | 0.1 |
| $V^{(4)}_{\alpha\alpha'}$ | Nb | Mo | Ta | W |
| Nb | −0.0 | −0.5 | 0.3 | 0.2 |
| Mo | −0.5 | 0.4 | −0.4 | 0.5 |
| Ta | 0.3 | −0.4 | 0.4 | −0.3 |
| W | 0.2 | 0.5 | −0.3 | −0.4 |
| $V^{(5)}_{\alpha\alpha'}$ | Nb | Mo | Ta | W |
| Nb | 0.5 | −0.9 | 0.9 | −0.5 |
| Mo | −0.9 | 2.0 | −2.1 | 1.0 |
| Ta | 0.9 | −2.1 | 2.2 | −1.0 |
| W | −0.5 | 1.0 | −1.0 | 0.5 |
| $V^{(6)}_{\alpha\alpha'}$ | Nb | Mo | Ta | W |
| Nb | −1.0 | 0.3 | −0.3 | 1.0 |
| Mo | 0.3 | −0.5 | 0.4 | −0.2 |
| Ta | −0.3 | 0.4 | −0.4 | 0.2 |
| W | 1.0 | −0.2 | 0.2 | −1.0 |

**Table C.3** Atom-atom pairwise interactions for the five-component VNbMoTaW refractory high-entropy alloy considered in Chap. 4, fitted from $S^{(2)}$s evaluated at $T = 1200$ K for the specified equiatomic composition. The lattice parameter used in the $S^{(2)}$ calculation is the LDA-optimised parameter of $a = 3.119$ Å. All interaction values are in units of meV. Interactions are dominated by the first two shells

| $V^{(1)}_{\alpha\alpha'}$ | V | Nb | Mo | Ta | W |
|---|---|---|---|---|---|
| V | −18.1 | 4.4 | −17.1 | 22.9 | 7.9 |
| Nb | 4.4 | 4.8 | −8.6 | 7.3 | −7.8 |
| Mo | −17.1 | −8.6 | 21.2 | −13.3 | 17.8 |
| Ta | 22.9 | 7.3 | −13.3 | 3.8 | −20.7 |
| W | 7.9 | −7.8 | 17.8 | −20.7 | 2.8 |
| $V^{(2)}_{\alpha\alpha'}$ | V | Nb | Mo | Ta | W |
| V | 54.2 | −3.2 | −4.3 | −17.6 | −29.1 |
| Nb | −3.2 | 4.4 | −3.9 | 5.4 | −2.6 |
| Mo | −4.3 | −3.9 | 2.4 | −0.8 | 6.7 |
| Ta | −17.6 | 5.4 | −0.8 | 8.0 | 5.0 |
| W | −29.1 | −2.6 | 6.7 | 5.0 | 20.0 |
| $V^{(3)}_{\alpha\alpha'}$ | V | Nb | Mo | Ta | W |
| V | 1.0 | −0.7 | 0.0 | −0.3 | 0.0 |
| Nb | −0.7 | 0.5 | −1.0 | 1.2 | 0.0 |
| Mo | 0.0 | −1.0 | 2.8 | −2.8 | 1.0 |
| Ta | −0.3 | 1.2 | −2.8 | 2.9 | −1.0 |
| W | 0.0 | 0.0 | 1.0 | −1.0 | −0.1 |
| $V^{(4)}_{\alpha\alpha'}$ | V | Nb | Mo | Ta | W |
| V | −8.4 | 0.3 | 2.2 | 1.4 | 4.4 |
| Nb | 0.3 | 0.2 | −0.5 | 0.3 | −0.3 |
| Mo | 2.2 | −0.5 | 0.1 | −1.1 | −0.8 |
| Ta | 1.4 | 0.3 | −1.1 | 0.5 | −1.1 |
| W | 4.4 | −0.3 | −0.8 | −1.1 | −2.2 |
| $V^{(5)}_{\alpha\alpha'}$ | V | Nb | Mo | Ta | W |
| V | 2.8 | 0.9 | −2.8 | 1.5 | −2.3 |
| Nb | 0.9 | −0.5 | −0.2 | −0.3 | 0.1 |
| Mo | −2.8 | −0.2 | 2.9 | −1.6 | 1.7 |
| Ta | 1.5 | −0.3 | −1.6 | 0.9 | −0.6 |
| W | −2.3 | 0.1 | 1.7 | −0.6 | 1.1 |
| $V^{(6)}_{\alpha\alpha'}$ | V | Nb | Mo | Ta | W |
| V | 1.9 | −1.6 | −0.8 | 0.1 | 0.4 |
| Nb | −1.6 | −0.3 | 0.9 | −0.1 | 1.2 |
| Mo | −0.8 | 0.9 | −0.4 | 0.7 | −0.3 |
| Ta | 0.1 | −0.1 | 0.7 | −0.6 | 0.0 |
| W | 0.4 | 1.2 | −0.3 | 0.0 | −1.3 |

**Table C.4** Atom-atom pairwise interactions for the NbMoTaW refractory high-entropy alloy considered in Chap. 5, fitted from $S^{(2)}$s evaluated at $T = 1200$ K for the specified composition. The lattice parameter used in the $S^{(2)}$ calculation is based on an experimental determination and set to $a = 3.226$ Å. All interaction values are in units of meV. Interactions are dominated by the first two shells

| $V^{(1)}_{\alpha\alpha'}$ | Nb | Mo | Ta | W |
|---|---|---|---|---|
| Nb | 7.4 | −11.9 | 11.7 | −7.2 |
| Mo | −11.9 | 10.5 | −11.5 | 12.9 |
| Ta | 11.7 | −11.5 | 12.2 | −12.4 |
| W | −7.2 | 12.9 | −12.4 | 6.6 |
| $V^{(2)}_{\alpha\alpha'}$ | Nb | Mo | Ta | W |
| Nb | 6.5 | −1.6 | 2.4 | −7.2 |
| Mo | −1.6 | 1.9 | −1.2 | 0.9 |
| Ta | 2.4 | −1.2 | 0.8 | −2.0 |
| W | −7.2 | 0.9 | −2.0 | 8.4 |
| $V^{(3)}_{\alpha\alpha'}$ | Nb | Mo | Ta | W |
| Nb | 0.2 | −1.0 | 0.9 | −0.1 |
| Mo | −1.0 | 2.2 | −2.1 | 0.9 |
| Ta | 0.9 | −2.1 | 2.0 | −0.8 |
| W | −0.1 | 0.9 | −0.8 | −0.0 |
| $V^{(4)}_{\alpha\alpha'}$ | Nb | Mo | Ta | W |
| Nb | −0.0 | −0.3 | 0.2 | 0.2 |
| Mo | −0.3 | 0.2 | −0.1 | 0.3 |
| Ta | 0.2 | −0.1 | 0.1 | −0.1 |
| W | 0.2 | 0.3 | −0.1 | −0.3 |
| $V^{(5)}_{\alpha\alpha'}$ | Nb | Mo | Ta | W |
| Nb | 0.1 | −0.6 | 0.7 | −0.1 |
| Mo | −0.6 | 1.5 | −1.6 | 0.7 |
| Ta | 0.7 | −1.6 | 1.6 | −0.7 |
| W | −0.1 | 0.7 | −0.7 | 0.1 |
| $V^{(6)}_{\alpha\alpha'}$ | Nb | Mo | Ta | W |
| Nb | −0.7 | 0.2 | −0.2 | 0.7 |
| Mo | 0.2 | −0.5 | 0.4 | −0.1 |
| Ta | −0.2 | 0.4 | −0.3 | 0.1 |
| W | 0.7 | −0.1 | 0.1 | −0.7 |

**Table C.5** Atom-atom pairwise interactions for the $Ti_{0.5}NbMoTaW$ refractory high-entropy alloy considered in Chap. 5, fitted from $S^{(2)}$s evaluated at $T = 1200$ K for the specified composition. The lattice parameter used in the $S^{(2)}$ calculation is based on an experimental determination and set to $a = 3.234$ Å. All interaction values are in units of meV. Interactions are dominated by the first two shells

| $V^{(1)}_{\alpha\alpha'}$ | Ti | Nb | Mo | Ta | W |
|---|---|---|---|---|---|
| Ti | −48.3 | 12.1 | −25.7 | 32.8 | 5.2 |
| Nb | 12.1 | 4.8 | −9.8 | 7.1 | −8.2 |
| Mo | −25.7 | −9.8 | 17.8 | −11.8 | 16.7 |
| Ta | 32.8 | 7.1 | −11.8 | 4.9 | −16.8 |
| W | 5.2 | −8.2 | 16.7 | −16.8 | 5.6 |
| $V^{(2)}_{\alpha\alpha'}$ | Ti | Nb | Mo | Ta | W |
| Ti | 78.4 | 5.2 | −4.0 | −10.4 | −30.2 |
| Nb | 5.2 | 3.1 | −2.9 | 1.8 | −4.6 |
| Mo | −4.0 | −2.9 | 1.6 | −0.6 | 3.9 |
| Ta | −10.4 | 1.8 | −0.6 | 2.3 | 1.8 |
| W | −30.2 | −4.6 | 3.9 | 1.8 | 14.1 |
| $V^{(3)}_{\alpha\alpha'}$ | Ti | Nb | Mo | Ta | W |
| Ti | −0.5 | −1.6 | 0.5 | −0.2 | 1.5 |
| Nb | −1.6 | 0.3 | −0.6 | 0.9 | 0.2 |
| Mo | 0.5 | −0.6 | 2.2 | −2.2 | 0.4 |
| Ta | −0.2 | 0.9 | −2.2 | 2.2 | −0.7 |
| W | 1.5 | 0.2 | 0.4 | −0.7 | −0.6 |
| $V^{(4)}_{\alpha\alpha'}$ | Ti | Nb | Mo | Ta | W |
| Ti | −11.8 | −0.6 | 2.1 | 0.4 | 4.1 |
| Nb | −0.6 | 0.2 | −0.2 | 0.3 | −0.0 |
| Mo | 2.1 | −0.2 | −0.1 | −0.3 | −0.3 |
| Ta | 0.4 | 0.3 | −0.3 | 0.2 | −0.5 |
| W | 4.1 | −0.0 | −0.3 | −0.5 | −1.2 |
| $V^{(5)}_{\alpha\alpha'}$ | Ti | Nb | Mo | Ta | W |
| Ti | −3.5 | 0.2 | −0.8 | 1.5 | 0.9 |
| Nb | 0.2 | −0.2 | −0.1 | 0.1 | 0.2 |
| Mo | −0.8 | −0.1 | 1.5 | −1.3 | 0.3 |
| Ta | 1.5 | 0.1 | −1.3 | 1.0 | −0.4 |
| W | 0.9 | 0.2 | 0.3 | −0.4 | −0.5 |
| $V^{(6)}_{\alpha\alpha'}$ | Ti | Nb | Mo | Ta | W |
| Ti | 9.3 | −1.1 | −2.8 | 0.8 | −1.5 |
| Nb | −1.1 | −0.4 | 0.4 | −0.1 | 0.7 |
| Mo | −2.8 | 0.4 | 0.2 | 0.3 | 0.5 |
| Ta | 0.8 | −0.1 | 0.3 | −0.4 | −0.2 |
| W | −1.5 | 0.7 | 0.5 | −0.2 | −0.2 |

**Table C.6** Atom-atom pairwise interactions for the TiNbMoTaW refractory high-entropy alloy considered in Chap. 5, fitted from $S^{(2)}$s evaluated at $T = 1200$ K for the specified composition. The lattice parameter used in the $S^{(2)}$ calculation is based on an experimental determination and set to $a = 3.240$ Å. All interaction values are in units of meV. Interactions are dominated by the first two shells

| $V^{(1)}_{\alpha\alpha'}$ | Ti | Nb | Mo | Ta | W |
|---|---|---|---|---|---|
| Ti | −46.6 | 16.1 | −18.2 | 36.3 | 12.3 |
| Nb | 16.1 | 1.6 | −9.0 | 1.9 | −10.6 |
| Mo | −18.2 | −9.0 | 22.2 | −13.0 | 18.0 |
| Ta | 36.3 | 1.9 | −13.0 | −3.0 | −22.2 |
| W | 12.3 | −10.6 | 18.0 | −22.2 | 2.5 |
| $V^{(2)}_{\alpha\alpha'}$ | Ti | Nb | Mo | Ta | W |
| Ti | 65.5 | −3.4 | −10.9 | −16.9 | −34.3 |
| Nb | −3.4 | 2.5 | −1.7 | 3.0 | −0.4 |
| Mo | −10.9 | −1.7 | 3.0 | 1.9 | 7.7 |
| Ta | −16.9 | 3.0 | 1.9 | 5.1 | 6.9 |
| W | −34.3 | −0.4 | 7.7 | 6.9 | 20.1 |
| $V^{(3)}_{\alpha\alpha'}$ | Ti | Nb | Mo | Ta | W |
| Ti | 0.1 | −2.0 | 0.7 | −0.5 | 1.8 |
| Nb | −2.0 | 0.7 | −0.2 | 1.0 | 0.5 |
| Mo | 0.7 | −0.2 | 1.9 | −2.2 | −0.2 |
| Ta | −0.5 | 1.0 | −2.2 | 2.3 | −0.6 |
| W | 1.8 | 0.5 | −0.2 | −0.6 | −1.4 |
| $V^{(4)}_{\alpha\alpha'}$ | Ti | Nb | Mo | Ta | W |
| Ti | −8.4 | 0.1 | 2.6 | 1.2 | 4.5 |
| Nb | 0.1 | 0.3 | −0.3 | 0.2 | −0.3 |
| Mo | 2.6 | −0.3 | −0.6 | −0.6 | −1.1 |
| Ta | 1.2 | 0.2 | −0.6 | 0.1 | −0.9 |
| W | 4.5 | −0.3 | −1.1 | −0.9 | −2.2 |
| $V^{(5)}_{\alpha\alpha'}$ | Ti | Nb | Mo | Ta | W |
| Ti | −4.5 | 0.0 | 1.1 | 1.0 | 2.4 |
| Nb | 0.0 | −0.2 | 0.1 | −0.1 | 0.2 |
| Mo | 1.1 | 0.1 | 0.9 | −1.3 | −0.7 |
| Ta | 1.0 | −0.1 | −1.3 | 0.8 | −0.4 |
| W | 2.4 | 0.2 | −0.7 | −0.4 | −1.5 |
| $V^{(6)}_{\alpha\alpha'}$ | Ti | Nb | Mo | Ta | W |
| Ti | 6.8 | −1.4 | −3.6 | 0.5 | −2.2 |
| Nb | −1.4 | −0.1 | 0.8 | −0.1 | 0.9 |
| Mo | −3.6 | 0.8 | 1.1 | 0.5 | 1.2 |
| Ta | 0.5 | −0.1 | 0.5 | −0.7 | −0.2 |
| W | −2.2 | 0.9 | 1.2 | −0.2 | 0.3 |

**Table C.7** Atom-atom pairwise interactions for the VNbMoTaW refractory high-entropy alloy considered in Chap. 5, fitted from $S^{(2)}$s evaluated at $T = 1200$ K for the specified composition. The lattice parameter used in the $S^{(2)}$ calculation is based on an experimental determination and set to $a = 3.183$ Å. All interaction values are in units of meV. Interactions are dominated by the first two shells

| $V^{(1)}_{\alpha\alpha'}$ | V | Nb | Mo | Ta | W |
|---|---|---|---|---|---|
| V | −28.0 | 5.4 | −14.0 | 24.2 | 12.4 |
| Nb | 5.4 | 5.7 | −8.5 | 6.7 | −9.3 |
| Mo | −14.0 | −8.5 | 19.1 | −12.5 | 16.0 |
| Ta | 24.2 | 6.7 | −12.5 | 2.4 | −20.9 |
| W | 12.4 | −9.3 | 16.0 | −20.9 | 1.7 |
| $V^{(2)}_{\alpha\alpha'}$ | V | Nb | Mo | Ta | W |
| V | 49.5 | −3.1 | −3.4 | −16.6 | −26.5 |
| Nb | −3.1 | 4.5 | −3.6 | 4.9 | −2.7 |
| Mo | −3.4 | −3.6 | 1.7 | −0.5 | 5.7 |
| Ta | −16.6 | 4.9 | −0.5 | 7.3 | 5.0 |
| W | −26.5 | −2.7 | 5.7 | 5.0 | 18.5 |
| $V^{(3)}_{\alpha\alpha'}$ | V | Nb | Mo | Ta | W |
| V | 1.5 | −0.9 | 0.3 | −0.8 | −0.0 |
| Nb | −0.9 | 0.6 | −1.0 | 1.4 | 0.0 |
| Mo | 0.3 | −1.0 | 2.5 | −2.7 | 0.9 |
| Ta | −0.8 | 1.4 | −2.7 | 3.0 | −0.8 |
| W | −0.0 | 0.0 | 0.9 | −0.8 | −0.1 |
| $V^{(4)}_{\alpha\alpha'}$ | V | Nb | Mo | Ta | W |
| V | −7.5 | 0.4 | 1.9 | 1.3 | 3.8 |
| Nb | 0.4 | 0.2 | −0.5 | 0.3 | −0.4 |
| Mo | 1.9 | −0.5 | 0.1 | −1.0 | −0.6 |
| Ta | 1.3 | 0.3 | −1.0 | 0.4 | −1.1 |
| W | 3.8 | −0.4 | −0.6 | −1.1 | −1.8 |
| $V^{(5)}_{\alpha\alpha'}$ | V | Nb | Mo | Ta | W |
| V | 2.2 | 0.6 | −2.1 | 1.0 | −1.7 |
| Nb | 0.6 | −0.5 | −0.2 | −0.1 | 0.3 |
| Mo | −2.1 | −0.2 | 2.4 | −1.4 | 1.3 |
| Ta | 1.0 | −0.1 | −1.4 | 1.0 | −0.4 |
| W | −1.7 | 0.3 | 1.3 | −0.4 | 0.6 |
| $V^{(6)}_{\alpha\alpha'}$ | V | Nb | Mo | Ta | W |
| V | 1.8 | −1.2 | −1.1 | 0.3 | 0.1 |
| Nb | −1.2 | −0.3 | 0.7 | −0.1 | 0.9 |
| Mo | −1.1 | 0.7 | −0.2 | 0.5 | −0.0 |
| Ta | 0.3 | −0.1 | 0.5 | −0.6 | −0.1 |
| W | 0.1 | 0.9 | −0.0 | −0.1 | −0.9 |

**Table C.8** Atom-atom pairwise interactions for the $Ti_{0.5}VNbMoTaW$ refractory high-entropy alloy considered in Chap. 5, fitted from $S^{(2)}$s evaluated at $T = 1200$ K for the specified composition. The lattice parameter used in the $S^{(2)}$ calculation is based on an experimental determination and set to $a = 3.197$ Å. All interaction values are in units of meV. Interactions are dominated by the first two shells

| $V^{(1)}_{\alpha\alpha'}$ | Ti | V | Nb | Mo | Ta | W |
|---|---|---|---|---|---|---|
| Ti | −30.2 | −35.8 | 20.8 | −19.8 | 40.9 | 8.9 |
| V | −35.8 | −29.6 | 7.7 | −5.8 | 25.5 | 19.8 |
| Nb | 20.8 | 7.7 | 1.9 | −8.6 | 1.0 | −12.3 |
| Mo | −19.8 | −5.8 | −8.6 | 22.4 | −14.6 | 16.4 |
| Ta | 40.9 | 25.5 | 1.0 | −14.6 | −5.7 | −26.5 |
| W | 8.9 | 19.8 | −12.3 | 16.4 | −26.5 | −1.8 |
| $V^{(2)}_{\alpha\alpha'}$ | Ti | V | Nb | Mo | Ta | W |
| Ti | 64.2 | 45.9 | −5.4 | −15.1 | −18.8 | −38.3 |
| V | 45.9 | 41.0 | −8.4 | −7.6 | −19.9 | −27.9 |
| Nb | −5.4 | −8.4 | 4.4 | −1.5 | 6.3 | 1.9 |
| Mo | −15.1 | −7.6 | −1.5 | 3.8 | 2.7 | 10.1 |
| Ta | −18.8 | −19.9 | 6.3 | 2.7 | 9.9 | 10.2 |
| W | −38.3 | −27.9 | 1.9 | 10.1 | 10.2 | 24.7 |
| $V^{(3)}_{\alpha\alpha'}$ | Ti | V | Nb | Mo | Ta | W |
| Ti | −0.1 | 1.8 | −2.1 | −0.4 | −0.2 | 1.0 |
| V | 1.8 | 3.1 | −1.8 | 0.4 | −2.1 | −0.6 |
| Nb | −2.1 | −1.8 | 1.0 | −0.4 | 1.6 | 0.6 |
| Mo | −0.4 | 0.4 | −0.4 | 2.3 | −2.4 | 0.4 |
| Ta | −0.2 | −2.1 | 1.6 | −2.4 | 3.3 | −0.4 |
| W | 1.0 | −0.6 | 0.6 | 0.4 | −0.4 | −0.5 |
| $V^{(4)}_{\alpha\alpha'}$ | Ti | V | Nb | Mo | Ta | W |
| Ti | −9.0 | −6.5 | 0.7 | 3.1 | 2.0 | 5.1 |
| V | −6.5 | −5.0 | 0.8 | 2.1 | 1.7 | 3.6 |
| Nb | 0.7 | 0.8 | 0.1 | −0.6 | 0.0 | −0.7 |
| Mo | 3.1 | 2.1 | −0.6 | −0.4 | −1.2 | −1.3 |
| Ta | 2.0 | 1.7 | 0.0 | −1.2 | 0.0 | −1.5 |
| W | 5.1 | 3.6 | −0.7 | −1.3 | −1.5 | −2.7 |
| $V^{(5)}_{\alpha\alpha'}$ | Ti | V | Nb | Mo | Ta | W |
| Ti | −4.5 | −0.2 | −0.3 | −0.1 | 1.2 | 1.7 |
| V | −0.2 | 0.9 | 0.2 | −0.8 | 0.4 | −0.6 |
| Nb | −0.3 | 0.2 | −0.4 | 0.1 | −0.2 | 0.4 |
| Mo | −0.1 | −0.8 | 0.1 | 1.8 | −1.3 | 0.3 |
| Ta | 1.2 | 0.4 | −0.2 | −1.3 | 0.9 | −0.3 |
| W | 1.7 | −0.6 | 0.4 | 0.3 | −0.3 | −0.6 |
| $V^{(6)}_{\alpha\alpha'}$ | Ti | V | Nb | Mo | Ta | W |
| Ti | 8.0 | 4.1 | −2.0 | −3.2 | −0.5 | −2.3 |
| V | 4.1 | 1.4 | −1.2 | −1.9 | 0.4 | −0.7 |
| Nb | −2.0 | −1.2 | 0.1 | 1.1 | −0.1 | 1.1 |
| Mo | −3.2 | −1.9 | 1.1 | 0.8 | 0.7 | 0.9 |
| Ta | −0.5 | 0.4 | −0.1 | 0.7 | −0.7 | −0.1 |
| W | −2.3 | −0.7 | 1.1 | 0.9 | −0.1 | −0.0 |

**Table C.9** Atom-atom pairwise interactions for the TiVNbMoTaW refractory high-entropy alloy considered in Chap. 5, fitted from $S^{(2)}$s evaluated at $T = 1200$ K for the specified composition. The lattice parameter used in the $S^{(2)}$ calculation is based on an experimental determination and set to $a = 3.209$ Å. All interaction values are in units of meV. Interactions are dominated by the first two shells

| $V^{(1)}_{\alpha\alpha'}$ | Ti | V | Nb | Mo | Ta | W |
|---|---|---|---|---|---|---|
| Ti | −29.6 | −36.6 | 23.3 | −13.4 | 42.5 | 14.4 |
| V | −36.6 | −29.3 | 10.5 | 1.8 | 27.6 | 26.9 |
| Nb | 23.3 | 10.5 | −2.6 | −9.7 | −5.6 | −16.5 |
| Mo | −13.4 | 1.8 | −9.7 | 24.0 | −17.5 | 15.1 |
| Ta | 42.5 | 27.6 | −5.6 | −17.5 | −14.8 | −33.2 |
| W | 14.4 | 26.9 | −16.5 | 15.1 | −33.2 | −7.1 |
| $V^{(2)}_{\alpha\alpha'}$ | Ti | V | Nb | Mo | Ta | W |
| Ti | 54.3 | 37.7 | −11.3 | −19.0 | −23.0 | −40.0 |
| V | 37.7 | 34.1 | −12.3 | −9.8 | −22.5 | −28.1 |
| Nb | −11.3 | −12.3 | 5.9 | 1.5 | 9.2 | 7.2 |
| Mo | −19.0 | −9.8 | 1.5 | 6.3 | 6.8 | 14.7 |
| Ta | −23.0 | −22.5 | 9.2 | 6.8 | 13.8 | 16.3 |
| W | −40.0 | −28.1 | 7.2 | 14.7 | 16.3 | 30.9 |
| $V^{(3)}_{\alpha\alpha'}$ | Ti | V | Nb | Mo | Ta | W |
| Ti | −0.2 | 2.0 | −2.5 | −0.1 | −0.6 | 1.3 |
| V | 2.0 | 3.4 | −2.3 | 0.4 | −2.8 | −0.8 |
| Nb | −2.5 | −2.3 | 1.6 | 0.2 | 2.0 | 1.1 |
| Mo | −0.1 | 0.4 | 0.2 | 1.9 | −2.1 | −0.3 |
| Ta | −0.6 | −2.8 | 2.0 | −2.1 | 3.6 | −0.0 |
| W | 1.3 | −0.8 | 1.1 | −0.3 | −0.0 | −1.3 |
| $V^{(4)}_{\alpha\alpha'}$ | Ti | V | Nb | Mo | Ta | W |
| Ti | −6.7 | −4.7 | 1.0 | 3.3 | 2.2 | 5.1 |
| V | −4.7 | −3.8 | 1.1 | 2.2 | 1.8 | 3.5 |
| Nb | 1.0 | 1.1 | −0.0 | −0.7 | −0.2 | −1.1 |
| Mo | 3.3 | 2.2 | −0.7 | −1.1 | −1.5 | −2.2 |
| Ta | 2.2 | 1.8 | −0.2 | −1.5 | −0.3 | −1.9 |
| W | 5.1 | 3.5 | −1.1 | −2.2 | −1.9 | −3.6 |
| $V^{(5)}_{\alpha\alpha'}$ | Ti | V | Nb | Mo | Ta | W |
| Ti | −5.1 | −0.6 | −0.0 | 1.6 | 1.2 | 3.1 |
| V | −0.6 | 0.7 | 0.1 | 0.0 | −0.1 | −0.2 |
| Nb | −0.0 | 0.1 | −0.2 | 0.1 | −0.2 | 0.2 |
| Mo | 1.6 | 0.0 | 0.1 | 0.7 | −1.4 | −1.0 |
| Ta | 1.2 | −0.1 | −0.2 | −1.4 | 0.9 | −0.4 |
| W | 3.1 | −0.2 | 0.2 | −1.0 | −0.4 | −1.9 |
| $V^{(6)}_{\alpha\alpha'}$ | Ti | V | Nb | Mo | Ta | W |
| Ti | 6.0 | 3.1 | −2.1 | −3.8 | −0.6 | −2.7 |
| V | 3.1 | 1.0 | −1.1 | −2.3 | 0.4 | −1.0 |
| Nb | −2.1 | −1.1 | 0.4 | 1.6 | 0.0 | 1.3 |
| Mo | −3.8 | −2.3 | 1.6 | 1.8 | 1.0 | 1.8 |
| Ta | −0.6 | 0.4 | 0.0 | 1.0 | −0.9 | 0.0 |
| W | −2.7 | −1.0 | 1.3 | 1.8 | 0.0 | 0.7 |

**Table C.10** Atom-atom interchange parameters for CrCoNi in a non-magnetic (or, equivalently, paramagnetic) state, computed at the experimental lattice spacing of 3.56 Å. All values in units of meV. Note that, for the CrCoNi system, the non-magnetic and paramagnetic states are identical; the DLM state collapses to a non-magnetic state in the self-consistency cycle

| $V^{(1)}_{\alpha\alpha'}$ | Ni | Co | Cr |
|---|---|---|---|
| Ni | −3.0 | 9.3 | −6.3 |
| Co | 9.3 | 9.2 | −18.4 |
| Cr | −6.3 | −18.4 | 24.7 |
| $V^{(2)}_{\alpha\alpha'}$ | Ni | Co | Cr |
| Ni | 3.6 | 0.4 | −4.0 |
| Co | 0.4 | 0.0 | −0.5 |
| Cr | −4.0 | −0.5 | 4.5 |
| $V^{(3)}_{\alpha\alpha'}$ | Ni | Co | Cr |
| Ni | 0.3 | 1.3 | −1.6 |
| Co | 1.3 | 0.7 | −2.0 |
| Cr | −1.6 | −2.0 | 3.7 |
| $V^{(4)}_{\alpha\alpha'}$ | Ni | Co | Cr |
| Ni | 3.3 | −0.1 | −3.2 |
| Co | −0.1 | −0.7 | 0.8 |
| Cr | −3.2 | 0.8 | 2.4 |

**Table C.11** Atom-atom interchange parameters for CrCoNi in a ferrimagnetic state, computed at the experimental lattice spacing of 3.56 Å. All values in units of meV

| $V^{(1)}_{\alpha\alpha'}$ | Ni | Co | Cr |
|---|---|---|---|
| Ni | −2.0 | 6.3 | −4.3 |
| Co | 6.3 | 3.4 | −9.6 |
| Cr | −4.3 | −9.6 | 13.8 |
| $V^{(2)}_{\alpha\alpha'}$ | Ni | Co | Cr |
| Ni | 3.7 | 0.3 | −4.0 |
| Co | 0.3 | 0.0 | −0.3 |
| Cr | −4.0 | −0.3 | 4.3 |
| $V^{(3)}_{\alpha\alpha'}$ | Ni | Co | Cr |
| Ni | 0.3 | 0.8 | −1.1 |
| Co | 0.8 | 0.2 | −1.0 |
| Cr | −1.1 | −1.0 | 2.0 |
| $V^{(4)}_{\alpha\alpha'}$ | Ni | Co | Cr |
| Ni | 2.8 | −0.0 | −2.8 |
| Co | −0.0 | 0.3 | −0.3 |
| Cr | −2.8 | −0.3 | 3.1 |

**Table C.12** Atom-atom interchange parameters for CrFeCoNi modelled in a non-magnetic state, computed at the experimental lattice spacing of 3.57 Å. All values in units of meV

| $V^{(1)}_{\alpha\alpha'}$ | Ni | Co | Fe | Cr |
|---|---|---|---|---|
| Ni | −8.2 | 5.3 | 9.9 | −7.1 |
| Co | 5.3 | 5.8 | 2.6 | −13.6 |
| Fe | 9.9 | 2.6 | −3.4 | −9.1 |
| Cr | −7.1 | −13.6 | −9.1 | 29.8 |
| $V^{(2)}_{\alpha\alpha'}$ | Ni | Co | Fe | Cr |
| Ni | 5.2 | 1.2 | −2.3 | −4.1 |
| Co | 1.2 | 0.3 | −0.2 | −1.3 |
| Fe | −2.3 | −0.2 | 1.6 | 0.9 |
| Cr | −4.1 | −1.3 | 0.9 | 4.5 |
| $V^{(3)}_{\alpha\alpha'}$ | Ni | Co | Fe | Cr |
| Ni | −0.4 | 1.0 | 1.0 | −1.7 |
| Co | 1.0 | 0.5 | −0.2 | −1.2 |
| Fe | 1.0 | −0.2 | −0.8 | −0.0 |
| Cr | −1.7 | −1.2 | −0.0 | 2.9 |
| $V^{(4)}_{\alpha\alpha'}$ | Ni | Co | Fe | Cr |
| Ni | 4.9 | 0.4 | −2.5 | −2.8 |
| Co | 0.4 | −0.4 | −0.6 | 0.6 |
| Fe | −2.5 | −0.6 | 1.0 | 2.1 |
| Cr | −2.8 | 0.6 | 2.1 | −0.0 |

**Table C.13** Atom-atom interchange parameters for CrFeCoNi modelled in a paramagnetic state, computed at the experimental lattice spacing of 3.57 Å. All values in units of meV

| $V_{\alpha\alpha'}^{(1)}$ | Ni | Co | Fe | Cr |
|---|---|---|---|---|
| Ni | −4.6 | 8.3 | 1.3 | −5.0 |
| Co | 8.3 | 8.9 | −0.7 | −16.5 |
| Fe | 1.3 | −0.7 | −0.3 | −0.4 |
| Cr | −5.0 | −16.5 | −0.4 | 21.9 |
| $V_{\alpha\alpha'}^{(2)}$ | Ni | Co | Fe | Cr |
| Ni | 4.3 | 0.8 | −0.8 | −4.3 |
| Co | 0.8 | 0.1 | −0.1 | −0.8 |
| Fe | −0.8 | −0.1 | 0.1 | 0.8 |
| Cr | −4.3 | −0.8 | 0.8 | 4.2 |
| $V_{\alpha\alpha'}^{(3)}$ | Ni | Co | Fe | Cr |
| Ni | 0.0 | 1.2 | 0.1 | −1.4 |
| Co | 1.2 | 0.7 | −0.1 | −1.8 |
| Fe | 0.1 | −0.1 | −0.0 | 0.0 |
| Cr | −1.4 | −1.8 | 0.0 | 3.2 |
| $V_{\alpha\alpha'}^{(4)}$ | Ni | Co | Fe | Cr |
| Ni | 3.5 | 0.1 | −0.4 | −3.1 |
| Co | 0.1 | −0.6 | −0.0 | 0.6 |
| Fe | −0.4 | −0.0 | 0.0 | 0.4 |
| Cr | −3.1 | 0.6 | 0.4 | 2.2 |

**Table C.14** Atom-atom interchange parameters for CrFeCoNi modelled in a ferrimagnetic state, computed at the experimental lattice spacing of 3.57 Å. All values in units of meV

| $V^{(1)}_{\alpha\alpha'}$ | Ni | Co | Fe | Cr |
|---|---|---|---|---|
| Ni | −0.4 | 4.9 | −1.9 | −2.6 |
| Co | 4.9 | 3.3 | −0.4 | −7.7 |
| Fe | −1.9 | −0.4 | 8.6 | −6.2 |
| Cr | −2.6 | −7.7 | −6.2 | 16.5 |
| $V^{(2)}_{\alpha\alpha'}$ | Ni | Co | Fe | Cr |
| Ni | 3.9 | 0.5 | −0.5 | −3.8 |
| Co | 0.5 | 0.2 | −1.2 | 0.6 |
| Fe | −0.5 | −1.2 | −2.2 | 3.9 |
| Cr | −3.8 | 0.6 | 3.9 | −0.6 |
| $V^{(3)}_{\alpha\alpha'}$ | Ni | Co | Fe | Cr |
| Ni | 0.2 | 0.4 | −0.6 | 0.0 |
| Co | 0.4 | 0.1 | 0.1 | −0.6 |
| Fe | −0.6 | 0.1 | 1.5 | −1.0 |
| Cr | 0.0 | −0.6 | −1.0 | 1.6 |
| $V^{(4)}_{\alpha\alpha'}$ | Ni | Co | Fe | Cr |
| Ni | 2.0 | 0.2 | 1.0 | −3.2 |
| Co | 0.2 | 0.5 | 0.7 | −1.4 |
| Fe | 1.0 | 0.7 | −0.8 | −0.9 |
| Cr | −3.2 | −1.4 | −0.9 | 5.5 |

**Table C.15** Atom-atom interchange parameters for CrMnFeCoNi modelled in a non-magnetic state, computed at the experimental lattice spacing of 3.59 Å. All values in units of meV

| $V_{\alpha\alpha'}^{(1)}$ | Ni | Co | Fe | Mn | Cr |
|---|---|---|---|---|---|
| Ni | −10.6 | 3.2 | 8.5 | 5.7 | −6.8 |
| Co | 3.2 | 5.6 | 4.0 | −1.5 | −11.3 |
| Fe | 8.5 | 4.0 | −0.3 | −4.5 | −7.7 |
| Mn | 5.7 | −1.5 | −4.5 | −3.2 | 3.5 |
| Cr | −6.8 | −11.3 | −7.7 | 3.5 | 22.3 |
| $V_{\alpha\alpha'}^{(2)}$ | Ni | Co | Fe | Mn | Cr |
| Ni | 7.3 | 2.5 | −1.6 | −3.9 | −4.3 |
| Co | 2.5 | 0.8 | −0.4 | −1.1 | −1.8 |
| Fe | −1.6 | −0.4 | 0.8 | 1.1 | 0.2 |
| Mn | −3.9 | −1.1 | 1.1 | 2.0 | 1.8 |
| Cr | −4.3 | −1.8 | 0.2 | 1.8 | 4.1 |
| $V_{\alpha\alpha'}^{(3)}$ | Ni | Co | Fe | Mn | Cr |
| Ni | −0.3 | 0.9 | 0.9 | 0.1 | −1.4 |
| Co | 0.9 | 0.5 | 0.0 | −0.5 | −0.9 |
| Fe | 0.9 | 0.0 | −0.4 | −0.4 | −0.1 |
| Mn | 0.1 | −0.5 | −0.4 | 0.1 | 0.8 |
| Cr | −1.4 | −0.9 | −0.1 | 0.8 | 1.7 |
| $V_{\alpha\alpha'}^{(4)}$ | Ni | Co | Fe | Mn | Cr |
| Ni | 5.8 | 1.1 | −1.9 | −2.9 | −2.0 |
| Co | 1.1 | −0.2 | −0.7 | −0.5 | 0.3 |
| Fe | −1.9 | −0.7 | 0.3 | 1.0 | 1.3 |
| Mn | −2.9 | −0.5 | 1.0 | 1.5 | 0.9 |
| Cr | −2.0 | 0.3 | 1.3 | 0.9 | −0.4 |

**Table C.16** Atom-atom interchange parameters for CrMnFeCoNi modelled in a paramagnetic state, computed at the experimental lattice spacing of 3.59 Å. All values in units of meV

| $V^{(1)}_{\alpha\alpha'}$ | Ni | Co | Fe | Mn | Cr |
|---|---|---|---|---|---|
| Ni | −6.3 | 7.2 | 1.3 | 2.5 | −4.7 |
| Co | 7.2 | 9.5 | 0.9 | −3.8 | −13.7 |
| Fe | 1.3 | 0.9 | 0.0 | −0.7 | −1.5 |
| Mn | 2.5 | −3.8 | −0.7 | −0.9 | 2.9 |
| Cr | −4.7 | −13.7 | −1.5 | 2.9 | 17.1 |
| $V^{(2)}_{\alpha\alpha'}$ | Ni | Co | Fe | Mn | Cr |
| Ni | 5.9 | 1.8 | −0.2 | −3.1 | −4.3 |
| Co | 1.8 | 0.5 | 0.0 | −1.0 | −1.3 |
| Fe | −0.2 | 0.0 | −0.0 | 0.1 | 0.1 |
| Mn | −3.1 | −1.0 | 0.1 | 1.6 | 2.3 |
| Cr | −4.3 | −1.3 | 0.1 | 2.3 | 3.2 |
| $V^{(3)}_{\alpha\alpha'}$ | Ni | Co | Fe | Mn | Cr |
| Ni | 0.1 | 1.3 | 0.2 | −0.1 | −1.4 |
| Co | 1.3 | 0.8 | 0.1 | −0.6 | −1.5 |
| Fe | 0.2 | 0.1 | 0.0 | −0.1 | −0.2 |
| Mn | −0.1 | −0.6 | −0.1 | 0.1 | 0.8 |
| Cr | −1.4 | −1.5 | −0.2 | 0.8 | 2.4 |
| $V^{(4)}_{\alpha\alpha'}$ | Ni | Co | Fe | Mn | Cr |
| Ni | 4.0 | 0.4 | −0.0 | −1.9 | −2.5 |
| Co | 0.4 | −0.6 | −0.0 | −0.1 | 0.3 |
| Fe | −0.0 | −0.0 | −0.0 | 0.0 | 0.1 |
| Mn | −1.9 | −0.1 | 0.0 | 0.9 | 1.1 |
| Cr | −2.5 | 0.3 | 0.1 | 1.1 | 1.0 |

**Table C.17** Atom-atom interchange parameters for CrMnFeCoNi modelled in a ferrimagnetic state, computed at the experimental lattice spacing of 3.59 Å. All values in units of meV

| $V_{\alpha\alpha'}^{(1)}$ | Ni | Co | Fe | Mn | Cr |
|---|---|---|---|---|---|
| Ni | −5.2 | 4.3 | −1.6 | 5.6 | −3.1 |
| Co | 4.3 | 4.5 | −1.0 | −0.5 | −7.3 |
| Fe | −1.6 | −1.0 | 6.3 | −6.0 | 2.2 |
| Mn | 5.6 | −0.5 | −6.0 | 3.0 | −2.1 |
| Cr | −3.1 | −7.3 | 2.2 | −2.1 | 10.3 |
| $V_{\alpha\alpha'}^{(2)}$ | Ni | Co | Fe | Mn | Cr |
| Ni | 6.2 | 1.8 | −0.1 | −3.5 | −4.4 |
| Co | 1.8 | 0.7 | −0.9 | −0.0 | −1.5 |
| Fe | −0.1 | −0.9 | −3.5 | 3.5 | 1.0 |
| Mn | −3.5 | −0.0 | 3.5 | −1.5 | 1.5 |
| Cr | −4.4 | −1.5 | 1.0 | 1.5 | 3.5 |
| $V_{\alpha\alpha'}^{(3)}$ | Ni | Co | Fe | Mn | Cr |
| Ni | 0.1 | 0.7 | −0.2 | 0.3 | −1.0 |
| Co | 0.7 | 0.5 | 0.5 | −0.8 | −0.9 |
| Fe | −0.2 | 0.5 | 1.8 | −1.7 | −0.4 |
| Mn | 0.3 | −0.8 | −1.7 | 1.5 | 0.7 |
| Cr | −1.0 | −0.9 | −0.4 | 0.7 | 1.6 |
| $V_{\alpha\alpha'}^{(4)}$ | Ni | Co | Fe | Mn | Cr |
| Ni | 3.7 | 0.6 | 0.8 | −2.7 | −2.4 |
| Co | 0.6 | 0.4 | 0.7 | −1.0 | −0.7 |
| Fe | 0.8 | 0.7 | −0.8 | 0.4 | −1.2 |
| Mn | −2.7 | −1.0 | 0.4 | 0.8 | 2.4 |
| Cr | −2.4 | −0.7 | −1.2 | 2.4 | 1.9 |

# Appendix D
# Code and Data Availability

## D.1 Code Availability

- At the time of writing, the all-electron *Hutsepot* code [1] used for self-consistent KKR-CPA calculations is available from https://hutsepot.jku.at/.
- The $S^{(2)}$ code for perturbatively analysing the CPA reference medium and evaluating the atomic short-range order was originally developed by Prof. Julie B. Staunton and is available from either her or the author upon reasonable request.
- The code developed by the author for performing the lattice-based Monte Carlo simulations of alloy phase behaviour is freely available at: https://zenodo.org/doi/10.5281/zenodo.10371917.
- The *MARMOT* code [2], used for calculations of magnetic properties at both zero and finite temperature, is available from https://warwick.ac.uk/marmotcode.

## D.2 Data Availability

- A repository containing the data associated with this study is freely available on Zenodo: https://doi.org/10.5281/zenodo.10972964.

## References


1. M. Hoffmann, A. Ernst, W. Hergert, V.N. Antonov, W.A. Adeagbo, R.M. Geilhufe, H. Ben Hamed, Phys. Status Solidi (b) **257**(7), 1900671 (2020). https://doi.org/10.1002/pssb.201900671
2. C.E. Patrick, J.B. Staunton, Electron. Struct. **4**(1), 017001 (2022). https://doi.org/10.1088/2516-1075/ac5755





C. D. Woodgate, *Modelling Atomic Arrangements in Multicomponent Alloys*, Springer Series in Materials Science 346, https://doi.org/10.1007/978-3-031-62021-8

# Index




C. D. Woodgate, *Modelling Atomic Arrangements in Multicomponent Alloys*, Springer Series in Materials Science 346,
https://doi.org/10.1007/978-3-031-62021-8

GPSR Compliance

*The European Union's (EU) General Product Safety Regulation (GPSR) is a set of rules that requires consumer products to be safe and our obligations to ensure this.*

*If you have any concerns about our products, you can contact us on ProductSafety@springernature.com*

In case Publisher is established outside the EU, the EU authorized representative is:

Springer Nature Customer Service Center GmbH
Europaplatz 3
69115 Heidelberg, Germany

**Batch number: 10732044**

Printed by Printforce, the Netherlands